70 CENTURIES to go for the universe

VOLUME 1: THE BEGINNING

Second edition

Phineas NZENG

Self-published.

Second edition: March 2026

To learn more about the author and future projects:

🌐 www.phineas.blog

✉ hello@phineas.blog

From His first death and resurrection, the worlds and those who dwell in them were founded; through His second death and resurrection, the worlds and their inhabitants were redeemed from death forever.

This is the command He received from the Father: He has the power to lay down His life at will, and the power to take it up again.

Contents

Foreword

Tell me, the stars your eyes sometimes gaze upon at night when the sky is clear — are they eternal?

That beautiful starry vault, which from early childhood to old age never ceases to amaze the earth-dwellers — will it one day vanish?

What do people often answer when asked how much time the universe has left to live?

And you — how would you answer that question?

And I — what would I say in response to it?

Indeed, the universe — this vast container of countless stars that humanity so deeply longs to see up close — is not eternal.

The most reliable of all sources, the one that foretold the fall of the Ottoman Empire, the Holocaust, and even the technological advances of the twenty-first century nearly two millennia in advance, has revealed to us that there will come a time when the universe shall cease to exist.

If the exact day and hour of its disappearance are unknown — for YHWH the Father has set them by His own authority — the century of its end, however, has been revealed to us.

So then, will it vanish 70 centuries after the birth of Yeshua?

No — much sooner than that! And that is what you are about to discover.

CHAPTER 1: THE KEYS

ACCORDING TO SCIENCE

According to our science books, in the beginning was the Big Bang...

No worries! Let's explain the big word (Big Bang): the Big Bang theory says that in the beginning, there was absolutely nothing — no time, no space, no matter, and therefore no universe.

So there was nothing at all, when suddenly, a particle smaller than anything known to be the smallest and most invisible on Earth exploded into a mixture of light, heat and dust, which immediately spread out over a distance almost as great as the actual universe, i.e. a distance that is beyond human comprehension; don't even think about calculating!

So, this mixture of light, heat, and dust eventually gave rise — after several billions of years — to more than 2000 billion galaxies, each containing around 100 billion stars (the sun is also a star in a galaxy), with each star having more or less planets orbiting around it, much like Earth and seven other planets orbit the sun.

Then, 9 billion years after the Big Bang, life was born on earth, first taking shape in the waters, before gradually migrating to dry land and evolving into Homo sapiens sapiens, a human being very much like us.

Did science get it wrong? Not necessarily! It simply hasn't said everything; because science is based on facts, and facts do not always say everything, nor do they always prove everything. That's why Earth will certainly continue to have researchers as long as it exists — and their presence in this world remains valuable.

ACCORDING TO THE WORD OF GOD, THE BIBLE

In the beginning — it was *God!*

What did God do? He created.

> ***In the beginning God created the heavens and the earth.*** (Genesis 1:1–2).

There it is! Our good old Genesis chapter 1! For years, I flat-out wondered if it wasn't just a myth about creation — until one day I moved from asking "Is this a myth?" to:

- "What if this story hadn't aged a single day?"
- "What if this account told of things that our telescopes and space probes might never fully uncover?

So, I dug a little deeper in my research — and you would be absolutely amazed by all that I discovered (not by myself, of course!) — it's just... wow!

I then decided to share these discoveries with you to help you understand what YHWH (God) has revealed to us about the beginning of the ages He created, so that you might better grasp their flow and their outcome.

However, we won't go any further until I first tell you about a few tips we'll be using to understand the **first chapter of the book of Genesis**.
Below are the keys:

Key 1

In **Genesis 1**, when a word that appeared in a previous verse is repeated, it generally no longer carries the same meaning as before.

For example, the words **"heaven"** and **"earth"** have very different meanings depending on which section of Genesis chapter 1 you're reading.

Key 2

In Genesis, the elements are created simply by the commands of YHWH (God): He declares or commands something, and at the time He has appointed, that thing comes into existence.

Whoever says "God's command" is also saying "God's Word." And there are some truths we should know about the Word of God:

- The Word of God creates and acts by Itself.

  For He spoke, and it came to be; He commanded, and it stood firm *(Psalm 33:9).*

- The Word of God can create things in two realms: a material and visible reality, and an immaterial and invisible reality.

  For the Word of God is alive and active. Sharper than any double-edged sword *(So the Word of God holds absolute power over two realms: it is a double-edged sword);* ***it penetrates even to dividing soul and spirit*** *(It acts on invisible realms: soul and spirit),* ***joints and marrow*** *(It also acts on visible realities: joints and marrow)... (Hebrews 4:12)* .

In creating the world, it pleased YHWH to make it in two realms — that is, one single world, but expressed in two distinct realities: a material and physical reality, and an immaterial and spiritual reality.

This is an important point to keep in mind in order to better understand the creation account recorded in Genesis 1 .

The two keys mentioned above are therefore tools we will keep in hand to help us better understand things as we move forward.

So, can we believe what the book of Genesis teaches about the origins of the world?
Answer: In fact, to gain a global understanding of the cosmos — given the limitations imposed by the laws of physics — humanity still has nothing else to date that offers a comprehensive view of the cosmos that surrounds us, except the biblical accounts... if, of course, they are understood.

Indeed, truly understanding Genesis 1 reveals that our current knowledge of the universe both begins and ends halfway — in terms of time and space.

Genesis 1 thus helps us better locate ourselves in both time and space.

For, aside from the way it reports chronology — which does not follow our human conventions — the account of the creation of the world and of humankind in the opening chapters of the book of Genesis is, in my view, the most accurate story ever told about:

- The events that took place before the birth of the universe;
- What lies beyond the universe;
- The birth of the universe, and its limits — which our exploration tools cannot cross;
- The birth of planets and stars, with Earth taken in isolation to tell how YHWH (God), progressively introduced life upon it.

Indeed, YHWH sometimes uses vocabulary that is uniquely His own. At times, what He calls "tree" is not always a tree as human conventions define it; what He calls "water" is not always water as we know it; and sometimes what He calls "day" is not necessarily a day as we understand it. And that's normal — nothing obliges Him to align Himself with our conventions!

[Bonus:
Faith and science are both gifts from YHWH to His conscious creatures, but sometimes science is a very slow autopsy of what could be known and obtained at once through faith.
That's why science should not forbid itself to believe.
For the one who insists on first understanding by what mechanism the Red Sea will part before striking it with his staff will likely have long died under the Egyptian's sword by the time he finds the answer to his question.
Faith is not absurd — it's simply beyond us!]

So, would you like to learn one or two more things about the world and humanity that you may not have known until now?
Things that are logical and explainable — not absurdities?
What follows will surely make you go "Wow!"

CHAPTER 2: THE THREE HEAVENS

Indeed, the account in Genesis 1 helps us understand that there are three heavens. By "heaven", you should understand three distinct spaces above our heads, each with its own contents:

- **The first heaven:** from a physical standpoint, it is the only one our physical eyes can see and the only one our space exploration instruments can access. And there is a reason for that — just be patient, we'll get to it soon!
  To sum it up: the first heaven is the entire cosmos surrounding planet Earth — that is: space, the sun, the moon, the stars, and the other planets *(see Diagram 1 below for more detail).*

- **The second heaven,** or the middle heaven, is a space beyond our universe — beyond the first heaven.
  Like our universe, the second heaven is certainly both a physical and immaterial space *(see Diagram 1 for more details)*.

- **The third heaven,** or the highest heaven, is the Heaven of heavens *(see Diagram 1)*, the place where YHWH has His throne.
  Certain clues suggest that the Heaven of heavens is also both physical and immaterial, and you'll understand this better when we discuss the origin of the dust of the world.

> ***But will God really dwell on earth? The heavens, even the highest heaven cannot contain You...*** *(1 Kings 8:27)*

DIAGRAM 1: THE EARTH AND THE THREE HEAVENS
Below is the lower part of a sphere (a globe) cut in half.

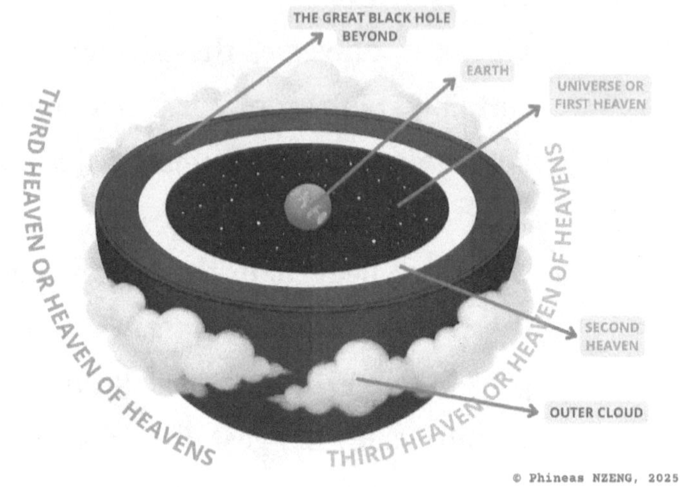

The three heavens described above, along with the earth we walk upon, were formed through a process and over a certain period of time. This duration is described in days according to YHWH's chronology, and in billions of years according to the results of our human research on the universe.

Indeed, YHWH (God) has His own way of describing and determining the time He created. He does not always call "day" what we call "day", nor "month" what we call "month". YHWH therefore does what He wants with the time He created, and wisdom would simply be to ask Him for an understandable conversion of time that our baby-like brains can understand.

It is therefore this process of the creation of the heavens and the earth — this kind of embryogenesis of creation — that **Genesis chapter 1** and other similar biblical accounts describe to us.

> *[Bonus*
>
> ***Elohim** = A Hebrew word literally translated as "Gods"; the singular form is Eloah (God).*
>
> ***YHWH** = The proper name of Elohim (God) in the Hebrew Bible. It is a tetragrammaton (a four-letter word), often transliterated as "Yahweh" or "Jehovah" in English.*
>
> *Although Elohim is actually a plural word, the Hebrew language always conjugates it in the singular ("For **Elohim knows** that when you eat from it..." — **Genesis 3:5**).*
>
> *YHWH, Elohim, is the Unified Being composed of the Principal Being — the Father — and His two*

thinking Arms (two Extensions, two Executors of designs), Who are:

*1.**His Word**, or Creative Substance, Who possesses His own consciousness and is a distinct Being in Himself.*

*2,**His Spirit** (His Thought), who is also a Being with His own consciousness — a distinct Being in Himself.*

They cannot contradict one another because They are One — one like the ever-watered Earth and the two Trees that grow from it, live through it, and manifest all the richness of it.

*And They can converse with one another because each has Their own consciousness: the ever-watered Earth (the Father) has a consciousness of His own, and so do His two Trees — each One **(See Zechariah 4:11–14, John 6:57, John 15:1)** .*

***Zechariah chapter 4:11–14** speaks of the Lord of all the universe and of the two Anointed Ones — the two Sons of Oil, or two Olive Trees — who stand before Him.*

A deeper study of ***Zechariah 4*** *reveals that one of the two Olive Trees is the One who lays the foundation of the house, that is, the foundation of the universe. He is called Zerubbabel, who is in fact a figure of Jesus Christ.*

It is also written that Zerubbabel (one of the olive trees, and the figure of ***Jesus****) necessarily needs the support of the other Olive Tree —* ***the Holy Spirit*** *— to fulfill the mission given by* ***the Father****.]*

CHAPTER 3: WATERS UNLIKE ANY OTHER

Genesis 1:1–2

"In the beginning God created the heavens and the earth. Now the earth was formless and empty, darkness was over the surface of the deep, and the Spirit of God was hovering over the waters."

Concerning the account of Genesis 1 , we previously stated that a word mentioned in an earlier verse almost never retains the same meaning in the following verses. Do you remember *Key 1*?

The "heavens" mentioned in the passage above ("In the beginning God created the **heavens** and the earth") most certainly refer to the Heaven of heavens, or the third heaven **(see Diagram 1 in the previous chapter)** and it is mentioned only once in

Genesis 1, without any description or further detail being associated with it.

And so, we move directly to the earth — and that's where things get even more interesting :

> ***Now the earth was formless and empty, darkness was over the surface of the deep, and the Spirit of God was hovering over the waters*** *(Genesis 1:2)*.

The "earth" mentioned in this second verse is not yet the planet Earth; it's not even close. This "earth" is actually the structure from which the second heaven and the universe (planet Earth + first heaven) were later formed.

When read in this version of the Bible (NIV), Genesis 1:2 might be difficult to understand for some, because certain terms may require clarification. For example, the terms:

- Formless and empty earth
- The deep (abyss)
- The waters

To better understand these terms, let us begin by revisiting Genesis 1:2 while keeping the original Hebrew term for "**formless and void**".

> ***Now the earth was tohu-bohu (Tohu wa-bohu), darkness was over the surface of the deep, and the Spirit of God was hovering over the waters** (Genesis 1:2).*

The term *tohu-bohu* (from the Hebrew *tohu wa-bohu*) means '**a state of extreme confusion and disorder.**' In other words, it refers to '**chaos, absolute desolation.**

A *tohu-bohu* earth, or one that is absolutely destroyed, simply means '**dust**'.

As for the term '***deep***', it means '**chasm**', '**precipice**', or '**emptiness**'.

Genesis 1:2 can therefore be reformulated as follows:

> *"The earth was in **a state of absolute chaos (= dust without shape)**; the chasm was covered with darkness, and the Holy Spirit was hovering*

over the face of the waters."

What serves as the container here is the chasm; therefore, the waters were certainly contained within the chasm, and within those waters was earth in a state of absolute destruction — that is, dust in suspension — certainly because something had reduced an earthly element to dust.
And all of it was covered in darkness, but at the surface of the waters contained in the chasm, the Holy Spirit was hovering."

Wow! Waters in a chasm of unfathomable immensity, covered in darkness? But wait a minute! Where did the chasm come from? Where and how was it carved out?

That's a good question, and the Bible does provide an answer.

The origin of the chasm: The chasm, or the deep, was created by YHWH, meaning that even the void did not exist before creation. Thus, through Solomon, Wisdom tells us:

> ***I was set up from everlasting, from the beginning, or ever the earth was.***
>
> ***When there were no depths, I was brought forth; when there were no fountains abounding with water*** *(Proverbs 8:23–24, King James Version–American Edition)*.

And we must understand that this chasm, this abyss, is not a dug-out hole, but rather a vast void created within "the absolute nothingness (even the void did not exist before creation)", and at the boundaries of which YHWH delineated a sphere (a globe), as it is written:

> ***I was there when He set the heavens in place, when He marked out the horizon on the face of the deep*** *(Proverbs 8:27)*.

Indeed, an horizon (a circle) drawn in a vast void—at a time when there was certainly not yet any notion of up or down, because there was certainly no gravity yet—must simply mean 'a sphere', that is, a circular boundary in all directions, and not just a flat circle as one might commonly understand it.

DIAGRAM 2: SPHERE MARKED OUT IN THE DEEP

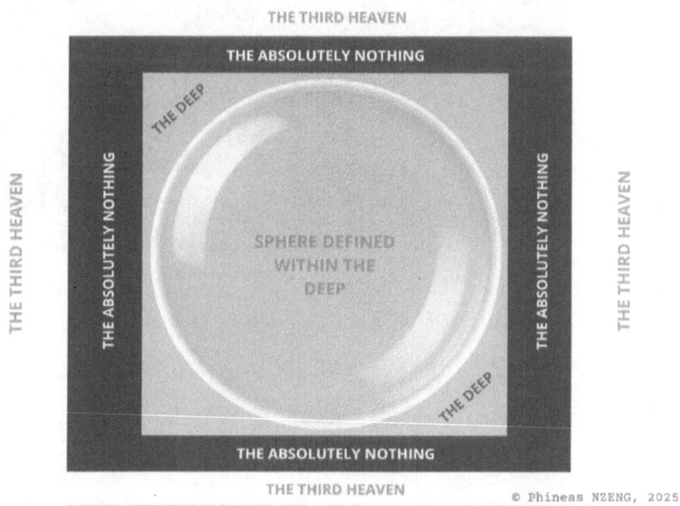

YHWH therefore marked out a spherical space at the boundaries of the great void (the deep), filled the spherical space with waters, and covered it all with darkness.

DIAGRAM 3: THE WATERS FILLING THE SPHERE MARKED OUT IN THE DEEP (EARLY STAGES)

In the circumstances as the Bible allows us to understand them, since gravity certainly did not yet exist, the sources from which the waters had sprung were surely suspended at the center of the sphere of the deep, from where they spread the waters in all directions at once, filling the sphere in a spherical manner, up to the boundaries of filling set by YHWH.

DIAGRAM 4: THE WATERS FILLING THE SPHERE MARKED OUT IN THE DEEP (LATER PHASE)

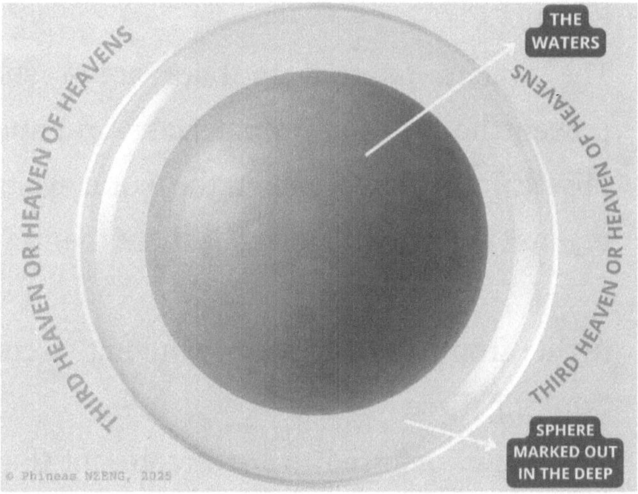

The waters...

Up to this point, we have:

- Waters in a chasm covered in darkness *(note: the darkness has not yet been illustrated up to this point)*;
- Dust suspended within the waters *(note: the dust has also not yet been illustrated)*;
- The Spirit of God hovering over the surface of the waters.

The waters mentioned in Genesis 1:2 are not of the same nature as the waters we consume on planet Earth.

The waters referred to here are a **substance present both in the immaterial world and in the physical world** (remember, the world exists in two realms), and we will now talk about them.

Right from the start, I wouldn't recommend drinking these waters.

The waters referred to here represent the sum of everything destructive that can be known; it is an environment where nothing can survive or remain whole. Everything disintegrates and turns to dust in these waters—except for the spirit (thought, reasoning).

In the physical universe, these waters would correspond to the constituent matter of black holes. And in the realm of the soul, these waters represent death (the realm of the dead, Sheol—a place where the soul disintegrates and returns to dust). More precisely, these waters are actually the first death, distinct from the second death, which is the lake of fire *(Revelation 21:8)* .

[Bonus: ***What is a black hole?***

Scientific research has taught us that there is a black hole at the center of the Milky Way, our galaxy.

Indeed, the Sun is a star, and planets orbit around it, including the Earth.

Most of the stars you see in the night sky are also like the Sun: they have planets orbiting around them.

A star and the planets that orbit around it together are called a ***stellar system*** *(from the Latin word "stella", meaning "star").*

A ***galaxy*** *is a very vast grouping of stars and their planets; there are usually hundreds of billions of stars in a single galaxy, and there are thought to be around* ***2 trillion observable galaxies****, with an unknown number of* *unobservable ones.*

All the galaxies together make up the **universe**. *The galaxy in which planet Earth and the solar system are located is called the* **Milky Way**.
At the center of the Milky Way—as at the center of all galaxies, in fact—is what we call a **black hole** *(see diagram 5 below).*

And to give you an idea: the **black hole** *at the center of the Milky Way is located about* **26,000 to 27,000 light-years** *from Earth.*
26,000 light-years means that if you were traveling at the speed of light (300,000 km/s), it would take you 26,000 years to reach the center of our galaxy.

Black holes *are places where nothing can remain intact: everything disintegrates and turns to dust.*
Science tells us that nothing can escape from a black hole—whatever enters it is reduced to dust and becomes trapped inside.

So, black holes can be seen as ***reservoirs of the waters of death****, but these are* ***physical waters of death****.]*

DIAGRAM 5: ILLUSTRATION OF A GALAXY, WITH A BLACK HOLE AT ITS CENTER

Indeed, we would certainly have known nothing about the waters and their nature if the Bible had not spoken of them.

David, likely referring to the sufferings that Jesus would later endure, but also to the suffering experienced by any soul that descends into the waters of death, wrote:

> *Save me, O God; for the waters are come in unto my soul.*
>
> *I sink in deep mire, where there is no standing : I am come into deep waters, where the floods overflow me.*
>
> *I'm weary of my crying: my throat is dried: mine eyes fail while I wait for my God...*
>
> *But as for me, my prayer is unto Thee, O Lord, in an acceptable time: O God, in the multitude of thy mercy hear me, in the truth of thy salvation.*
>
> *Deliver me out of the mire, and let me not sink: let me be delivered from them that hate me, and out of the deep waters.*
>
> *Let not the waterflood overflow me, neither let the deep swallow me up, and let not the pit shut her mouth upon me (Psalm 69:1–3 & 13–15, King James Version-American Edition).*

And elsewhere, on a similar topic, David says:

I'm poured out like water, and all my bones are out of joint: my heart is like wax; it is melted in the midst of my bowels.
My strength is dried up like a postherd, and my tongue cleaveth to my jaws; and thou hast brought me into the dust of death *(Psalm 22:14–15, King James Version-American Edition)*.

So that is the nature of the waters spoken of in Genesis 1:2: they reduce everything to dust, whether material or immaterial.
You see now why this is a kind of water you would definitely not want to drink, right?

These waters reduce both souls and bodies to dust ("you are dust, and to dust you shall return"), and at the foundation of the world, the Forerunner had to be reduced to dust in the waters. And from that dust, the universe and those who inhabit it were formed — something we will explore shortly.

Speaking of dust, remember this again: the world is dual in nature — it has both a **material** and an **immaterial** component.

As for the immaterial world, we must understand that the **dust** from which both our souls and the entire immaterial part of the cosmos were formed was, in the beginning, **suspended in the waters of death** (see *Psalm 69:1–3 & 13–16* and *Psalm 22:14–15* for references to the waters of death).

And as for the beginning of the material, physical world, we must understand that the **dust of our physical universe**—the dust of all galaxies, stars, planets, and everything they contain—was, in the beginning, suspended within the substance that makes up what we know in the universe as black holes (see the explanation above on black holes).
It was, therefore, the chaos, the *tohu-bohu*.

There you have it! Now that you understand Genesis 1 a little more clearly, we'll add some more about the formation of the deep and the origin of the waters, and we'll see how the dust ended up suspended in the waters contained in the deep...

CHAPTER 4: OF THE ABYSS AND THE WATERS

Up to this point, in speaking about the beginning, we have identified three key elements:

- The waters (of death) within a deep covered in darkness;
- The earth, reduced to chaos—that is, formless dust suspended in the waters;
- The Spirit of God hovering over the face of the waters.

These three elements made up the state of the **primitive Earth**, created by Elohim in the beginning. It is from this primitive Earth that He formed the second heaven, the Earth, and our cosmos, which is the first heaven.

About the Abyss...

As we saw earlier, the abyss is defined as "a gulf or a hole that is immeasurably vast and deep." It's also called "the deep".

Indeed, for us humans—who walk upon a land where we are held in place by gravity—we understand a hole as a space that has been dug out and bounded within the ground.

For someone digging a hole at the North Pole, it would be a hole going from the top to the bottom of the globe (from north to south). But for someone digging at the South Pole, the hole would go from the bottom to the top of the globe (from south to north).

Likewise, when someone standing at the South Pole looks up at the sky, they are actually looking downward in relation to the globe; and when someone standing on the equator looks up at the sky, they are in fact looking sideways.

So, when it comes to the beginning, the term "abyss" should not be understood as a hole dug into the ground, because—apart from Elohim—nothing at all existed at that moment: no space, no matter, no time — not even emptiness.

There was no gravity either to define what is up and what is down; there was absolutely nothing.

> ***I was set up from everlasting, from the beginning, or ever the earth was.***
> ***When there were no depths, I was brought forth; when there were no fountains abounding with water*** *(Proverbs 8:23-24, King James Version-American Edition)*.

Then Elohim created the Heaven (the third heaven) and the void — a vast emptiness which He marked out with a circle.
And since we are speaking of a context where there is no left or right, no up or down, by *circle* we must understand it to mean **a sphere**, a **globe**.

Have you guessed the name of that great void in which a sphere was marked out?
That void — was the abyss.

And so, Wisdom, speaking through Solomon, tells us:

> ***I was there when He set the heavens in place, when He marked out the horizon on the face of the deep, when He established the clouds above*** *(clouds different from the ones we see on Earth),* ***and fixed securely the fountains of the deep, when He gave the sea its boundary, so the waters would not overstep His command, and when He marked out the foundations of the earth*** (Proverbs 8:27-29) .

There you have it! At this stage, we now had Heaven (the third heaven) and a vast spherical void — the abyss.

Let us not forget that this void did not exist until it was created by YHWH, because even emptiness itself is a created entity. Before its creation, it did not exist; and after it was created, it began to exist.

So, if this void (the abyss) was created in the midst of absolute nothingness, then after its creation, wasn't it like a ball or a spherical pocket of emptiness stuck in the "nothing at all"? Wasn't it somewhat like a soccer ball completely buried and

pressed into the ground — the ball representing the abyss, and the ground all around it representing the "nothing at all"?
In that sense, it truly was an **abyss** — a vast **chasm**, a **great hole (see Diagram 2 in the previous chapter)**.

Then, after that, Elohim covered the walls of the abyss with clouds, and suspended the waters of death within the abyss, likely in a spherical manner.
That is to say, YHWH filled the abyss with water up to a certain limit, a limit set by His decree, by His command.

He filled it—not like one would fill a balloon on Earth, where gravity acts—but rather, He likely allowed the waters to spread out in a spherical shape within the abyss, up to the boundary defined by His decree, a boundary that acted like "gates" the waters could not pass **(see Diagram 3 and diagram 4 in the previous chapter)**.
And He wrapped the waters in darkness.

Who shut up the sea behind doors when it burst forth from the womb, when I made clouds its garment and wrapped it in thick darkness, when I fixed limits for it and set its doors and bars in place, when I said, 'This far you may come and no farther; here is where your proud waves halt (Job 38:8-11) .

DIAGRAM 6: WRAPPING OF THE WATERS BY THE CLOUD AND THE DARKNESS

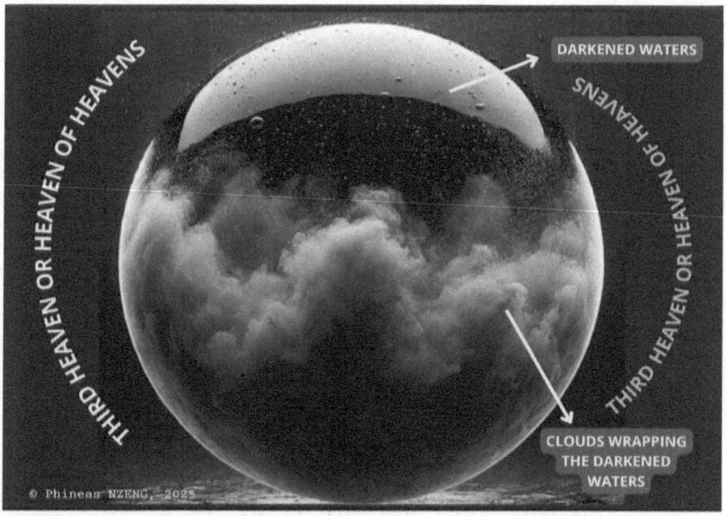

Illustration of the wrapping of the waters by darkness, and their covering by the cloud.

The outer boundary of the cloud is also the boundary of the sphere of the abyss; beyond that limit lies the third heaven,

or the heaven of heavens.
Here, the cloud only covers half of the sphere to allow us to see the waters beneath.

So, to summarize:

Out of absolute nothingness, YHWH carved out an abyss, upon whose boundaries He traced out a sphere, whose walls He covered with clouds.

He allowed the waters of death to fill the abyss up to the limit established by His decrees **(see Diagram 3 and diagram 4 in the previous chapter)**, and then He wrapped the waters of death in thick darkness.

So then, where did the waters of death come from?

Although the Bible does not explicitly say something like, *"Elohim created the waters..."*, it is certain that He is their Creator, for Wisdom lists them among His most ancient works.

> ***The Lord possessed me in the beginning of His way, before His works of old.***

> ***I was set up from everlasting, from the beginning, or ever the earth was.***
> ***When there were no depths, I was brought forth; when there were no fountains abounding with water*** *(Proverbs 8:22-24, King James Version-American Edition)*.

And also, wisdom teaches us that it is YHWH who securely fixed the fountains of the deep.

> ***I was there when He set the heavens in place, when He marked out the horizon on the face of the deep, when He established the clouds above*** *(clouds different from the ones we see on Earth),* ***and fixed securely the fountains of the deep, when He gave the sea its boundary, so the waters would not overstep His command, and when He marked out the foundations of the earth*** (Proverbs 8:27-29).

In addition, the Bible teaches us that YHWH has absolute power over the waters of death: the power to permit them to burst forth from within, the power to command and control them at will by His

Word (His decree, His order), and the power to dry them up forever.

- *Who shut up the sea behind doors when it burst forth from the womb, when I made clouds its garment and wrapped it in thick darkness, when I fixed limits for it and set its doors and bars in place, when I said, 'This far you may come and no farther; here is where your proud waves halt* *(Job 38:8-11)*.

- *Who says to the watery deep, 'Be dry and I will dry up your streams...* *(Isaiah 44:27)*

- *Then I saw "a new heaven and a new earth," for the first heaven and the first hearth had passed away, and there was no longer any sea* *(Revelation 21:1)*.

Thus, for all eternity, YHWH alone holds the power to bring back both souls and bodies from death. And whoever remains in communion with Him through His Arms—Jesus Christ and the Holy Spirit—cannot be held captive by the waters of death; resurrection and life are their inheritance forever.

***- The Lord brings death and makes alive; He brings down to the grave and raises up** (1 Samuel 2:6).*

- Jesus said to her, "I'm the resurrection and the life. He who believes in Me, though he may die, he shall live.
***And whoever lives and believes Me shall never die. Do you believe this** (John 11:25-26, NKJV)* ?

[Bonus:
The physical waters (black holes) are far removed from us, but the immaterial waters have their rivers and streams flowing among human souls, and they manifest as disturbances within our physical world.

These rivers and streams of the abyss, which dwell among us, are sin, disease, accidents, magical practices, witchcraft, and so on — in short, evil.

And every day on Earth, souls are swept away by these rivers and streams from the abyss, carried toward the great waters of the abyss (the sea) to be reduced to dust (for dust you are, and to dust you shall return), to be tormented and held captive, awaiting the day when the waters will give up their dead ***(Revelation 20:13)****. Not every so-called natural cause... is truly natural.*

But the good news is that those whose souls have been regenerated by God through Jesus Christ—those who have received Jesus and remain in communion with God (through obedience, prayer, Bible reading, praise, and regular worship)—have the power to dry up the rivers and streams directed against them.

And even if their souls fall into the waters, the resurrection that is in Jesus-Christ and YHWH the Father will bring them back, for they are not meant to wait for the resurrection of the dead in disintegration, torment, or the prison of the waters ***(Revelation 14:13).***

> ***- They will pass through the sea of trouble, the surging sea will be subdued and all the depths of the Nile will dry up. Assyria's pride will be brought down and Egypt's scepter will pass away.***
> ***I will strengthen them in the Lord and in His name they will live securely, declares the Lord (Zachariah 10:11-12).***
>
> ***- ... I will build my Church, and the gates of the Hades shall not prevail against it (Matthieu 16:18, NKJV).***
>
> ***- Jesus said to her, "I'm the resurrection and the life. He who believes in Me, though he may die, he shall live.And whoever lives and believes Me shall never die. Do you believe this (John 11:25-26, NKJV)?]***

Until now, what we have described is both a material and immaterial representation of things. So for the moment, we must consider all the elements of the creation that have been mentioned as simultaneously physical and immaterial.

However, we must also understand that time and distance do not flow the same way in both realities: the immaterial world—the world of our souls and dreams—is a mystery to us, and it follows its own laws.

> ***[A brief word of advice about the immaterial world:***
> *Anyone who belongs to Elohim does not look into the soul-world (the world of our dreams), nor enter it on their own initiative—even if, unfortunately, they had learned to do so in the past.*
> *One must wait for the Holy Spirit to carry them there! For this is exactly what the Bible reveals to us regarding all out-of-body experiences and all visions of the soul-world*

and the world of spirits.

And this is, in fact, the key difference between a prophet of God and a sorcerer: the sorcerer, sadly, knows how to gaze into the other world by his own means. But just like a smoker who quits smoking forever once he has truly accepted and received Jesus, the sorcerer who has accepted and received the Lord Jesus Christ should likewise forever stop entering, accessing or turning his eyes unto the immaterial world on his own; It 's first of all a matter of personal will and effort.

If he does not stop doing this, there will never be deliverance or salvation for him. For by looking into that world on his own, he shuts the door to the Holy Spirit—like a horse that wants to lead itself—and he continually keeps open the doors of his soul to contaminated breaths (evil spirits or impure intelligences), which will manipulate him, distort his judgment, and lead him down the wrong path.

As a result, sin and addictions will always have the upper hand over him. For it is the Holy Spirit within us who drives away evil spirits and helps us live according to the will of God. Without Him, the battle is lost from the start.

A spirit is a breath, a thought, an intelligence, and he can create spiritual scenarios, visions, in a human being. You only see him once he has influenced your own thoughts (your spirit) to make you see him—or when the Spirit of God has conditioned you to perceive him. He can therefore act without your awareness—and in fact, that is what most often happens.

Angels and contaminated breaths (demons) are spirits (**Psalm 104:4**)*.]*

That's it! Now that you know a bit more, let's talk about the dust (chaos) that was in the beginning in the waters: where did it come from? And how did we get from absolute chaos to a structured, well-formed world?

CHAPTER 5: THE CHAOS

A quick return to school...

The human body, even that of an animal, is made up of **molecules**, such as **proteins, fats, vitamins, hormones**, and so on. These molecules serve to build the body and also ensure its proper functioning.

Molecules themselves are made of even smaller units called atoms. An atom is, for example, **Carbon (C)**, **Hydrogen (H)**, or **Nitrogen (N)**, just to name a few.

A chaos within our bodies...

One of the most brilliant and vital processes that takes place in our body is digestion. It's something that happens every day, often without much thought, to both you and me—and even to animals.

For example, when a person eats chicken, the various chemical substances found in the fluids of their stomach and intestines break that chicken down into molecules, and then into atoms (see definitions above), which are absorbed into the bloodstream.

Once in the blood, these atoms are reassembled by certain organs into different molecules that are useful for building the human body and ensuring its proper functioning.

Indeed, a chicken has its own molecules: proteins, vitamins, fats, hormones, etc., but these are chicken molecules—and they cannot function properly in a human body. However, the atoms that make up the chicken's molecules are the same atoms—carbon, hydrogen, and so on—that are found in a human being.

That's why the human body must break down the chicken's molecules (fats, proteins, etc.) into individual atoms (carbon, hydrogen, etc.), so it can then rebuild its own molecules using those atoms.

It's a bit like demolishing a house and then reusing its bricks to build a completely different model of house.
And the blueprint that the body follows to reassemble those atoms into our own molecules has been written into our bodies from the moment of conception in the form of a lifelong program: our DNA.

In this sense, our stomachs and intestines are true places of chaos: the food we throw into them cannot remain whole in the liquid they contain, because that liquid is filled with chemical substances that break the food down and dissolve it.
Any food we place there is reduced to chaos, to dust, to the state of atoms.

Code in everything that lives...

Indeed, YHWH is a Planner, a Programmer by nature; He is not an interventionist in the way many might imagine. Most of the things He does, He programs and plans them in advance. That may be why miracles are rare events, and also why the answers to our prayers do not always appear immediately.

Almost everything YHWH does is first programmed and strategically timed, in order to come to pass at the right moment. For example, if you didn't take the time to plant during the rainy season, you may spend the entire dry season praying for rain, but the answer to your prayers may only manifest months later, in the next rainy season... unless there's a miracle!

For example, a little 4-year-old boy who starts praying to grow a beard will likely continue praying for the next 14 years of his life before seeing the answer to his prayer come to pass—unless a miracle occurs.

Yes, almost everything Elohim does is first programmed, just like an architect who completes the entire blueprint of a house on paper before beginning construction.

> ***The plans of the diligent lead to profit as surely as haste lead to poverty*** *(Proverbs 21:5).*

In the same way, for all living beings—microbes, plants, animals, and humans—the entire development and functioning of their organisms is already written into them from the very moment they are conceived: their size and weight at different ages, their skin color, hair color, the way their bodies are to reassemble atoms from digestion into molecules, and even part of their behavior—all of this is prewritten within them. The rest of their physical, and even behavioral development, is simply the execution of a prewritten plan, and that plan is the genetic code: DNA.

Wow, can you imagine, humanly speaking, how enormous the task would be without DNA? What if every small change that an organism needed to undergo required a specific command from YHWH?

- *If, for example, YHWH had to start commanding the first tooth of every child in the world to grow when the right time comes for each of them?*
- *Or if, after each digestion, He had to command the organisms, each according to the timing of its feeding, to form molecules with the atoms obtained?*

Of course, that would certainly not have been too much for YHWH, because, for example, ever since infection entered creation, He has often operated in an interventionist mode for certain affected matters, for the well-being of each one of His creatures. But in the beginning, that was not the case: Elohim had programmed everything to function on its own, forever ***(...Elohim rested on the seventh day from all His work which He had done...*** *Genesis 2:2****)***.

At the beginning of the world: the chaos...

Earlier, as we were describing the waters mentioned in Genesis 1 , we saw that they were waters of death: nothing thrown into them remains whole; they reduce everything to dust, to chaos—a bit like what happens when we digest food. Do you remember what happens to food in the liquid inside our stomachs and intestines?

And since we previously saw that the universe is both material and immaterial, we also understood that the waters mentioned in Genesis 1:2 are likewise both material and immaterial. The material waters would be the extremely destructive substance contained within the black holes of the universe, while the immaterial waters are a realm where souls are reduced to dust, to chaos—even though the spirit (that is, consciousness) connected to the soul remains intact and can feel the soul's suffering. And this is what the death described in certain Bible passages refers to—such as Psalm 22:14–16 and Psalm 69:1–3 & 13–16 .

We also said that the chaos (*tohu va bohu*) described in **Genesis 1:2** was due to the fact that a primordial element had been reduced to chaos, to dust (to atoms?) in the waters of death at the foundation of the world—much like food is reduced in our stomachs and intestines. And this is what we will now examine in detail:

Revelation 13:8 NKJV: ***All who dwell on the earth will worship him, whose names have not been written in the Book of Life of the Lamb slain from the foundation of the world.***

There it is! The Lamb slain from the foundation of the world!

We would certainly know very little about the dust through which Elohim founded the world,if this passage of the Bible had not been revealed to us.

The Lamb mentioned in this passage is the One called the Logos, or the Word, the Creative Substance who has consciousness (John 1:1–4 **and**

Revelation 19:11–16).

He is also one of the two Anointed Ones who stand before the Elohim of all the universe (Zechariah 4) , and is referred to as the Anointed One, the Ruler **(Daniel 9:25)**. He is also called the Arm of the Lord (Isaiah 53:1).

The Lamb of God—who was slain from the foundation of the world—is Jesus Christ (John 1:29–30) , the One from whom every atom of our souls and of the realm of souls has its origin, and also the One from whom every atom of our bodies and of the universe itself originates.

> ***All things were made through Him, and without Him nothing was made that was made*** *(John 1:3, NKJV)*.

Indeed, the Bible tells us of two successive creations of the world by YHWH . The creation of the first world, the one described in the book of Genesis, is already completely finished (*God ended His work which He had done... Genesis 2:2*), and the

creation of the second (final) world, the one described in Revelation, has already begun.
As you read this work, the destruction of the first world has also already begun, even as the construction of the second world is already underway.

With each creation—whether of the first world or the second—the Lamb had to be sacrificed in the waters first (***unless a grain of wheat falls into the ground and dies, it remains alone...*** *John 12:24*). So, from the moment we saw the Lamb among us, announcing His death and resurrection, the creation of a new world, a final and eternal world, was about to be triggered, forever.

> ***From that time on Jesus began to preach, "Repent for the Kingdom of Heaven has come near*** *(Matthew 4:17)* ***."***

Going through destruction is the most tragic experience one can have. Jesus knew it—because He had already experienced it (***"My Father, if it is possible, let this cup pass from Me..."*** *Matthew 26:39*).

Yet passing through destruction is also the ultimate proof of love. It is therefore out of love that the world was born, and it is out of love that it is now being reborn. For Elohim could have remained alone for all eternity if He had wished, but it pleased Him to create other beings to share in His happiness; that food you enjoy, for example—isn't that a form of happiness?

And when, because of human disobedience, the universe became infected—when the waters of death and wickedness found an entry point into the human soul, like seawater flooding a leaking boat—Elohim could have simply let us all sink into the waters, then cast us into the fire of hell with the rebellious angels forever.

However, for creating us, as well as for restoring us and recreating the world, Elohim sacrificed the Lamb to the waters of death, much like a man cutting off his own arm—except that this Arm had its own consciousness and feelings—and throwing it into the fire, so that worlds might (re)be born and live through it.

- ***Greater love has no one than this : to lay down one's life for one's friend*** *(John 15:13).*

- ***So God so loved the world that He gave His Only begotten Son, that whosoever believes in Him should not perish but have everlasting life*** *(John 3:16, NKJV).*

Yes, just as food is torn apart, broken down, and reduced to dust in the aggressive liquid of our digestive system, so too was the Lamb of God reduced to dust (to the state of atoms?) in the waters of death at the foundation of the world.
Thus, the entire future universe had just been founded, but it was first existing only as dust within the waters: it was **chaos, *tohu wa-bohu*, confusion.**

[Bonus

Someone might wonder: if the Lamb had been sacrificed, and if the dust resulting from His destruction by the waters had enabled YHWH to create the world, then how is it that the Lamb still existed, even to the point of one day taking on a human body on Earth, to announce the new world and to enter the waters once again?

Indeed, there are **two main things** *we must understand:*

*- **First:***

When it comes to a soul, the waters never destroy everything completely; there is always a kind of core that remains, for Jesus, like for the human souls, that all come from Him (grow out of Him... **John 15:5***), because they all originate from the Father, who lives forever.*

Thus, In the vision of **Ezekiel 37** *, for example, the dead souls are revealed to Ezekiel as dry bones in a valley, because something remained after everything else had disintegrated.*

And when, in **Zechariah 3** *, Jesus is brought out of the waters during His resurrection and* *appears wearing filthy garments (the sins of the world) before the Father, who then gives Him new garments, the Father compares Him to a firebrand snatched from the fire—that is, a piece of wood that had not completely burned, but was pulled out of the fire (the waters of death).*

We, like Jesus, are pieces of wood that cannot be completely burned or disintegrated.

> ***Therefore, as the fire devours the stubble, and the flame consumes the chaff, so their root will be as rottenness*** *(their source part will just rot, but not disintegrate),* ***and their blossom will ascend like dust*** *(the rest of their soul will fall apart);* ***because they have rejected the law of the Lord of hosts, and despised the law of the Holy one of Israel (Isaiah 5:24, NKJV).***

- Second:

The command received from the Father brings Jesus back from the waters and restores to Him all that He had left there.

For if God has a plan for what Jesus leaves behind in death (a plan to use it to create or recreate worlds), He also has a plan for Jesus Himself—and that plan is the Father's command that lives within Jesus (a kind of DNA?), the command to return from death and to recover His original state.

> ***Therefore My Father loves Me, because I lay down My life that I may take it again.***
>
> ***None takes it from Me, but I lay it down of Myself. I have power to lay it down, and I have power to take it again. This command I have received from My Father (John 10:17-18).]***

The program...

Just as there is always a program (our genetic code) that enables our bodies to reassemble digested atoms into our own molecules, likewise, it was not without a prewritten and established plan that YHWH dissolved His own Arm—that is, the Lamb—in the waters of death.

A plan for the reconstitution of the dust of His Arm into the universe had surely been prepared and written long beforehand, and that's what we're now going to see...

CHAPTER 6: THE DNA OF THE UNIVERSE

As we saw earlier, during the creation of the world, YHWH (God) cast into the waters of death His Thinking Arm (*Jesus-Christ, the Lamb of God*), so that the dust resulting from the part of Him that could be disintegrated by the waters would serve to create the universe.

We also previously said that Elohim (God) is a Planner, a Programmer; most of the things He does are planned long in advance, and when we begin to see them unfold, it is often the execution of a pre-established plan, like a machine programmed ahead of time, which simply begins to operate on its own.

- ***See the former things have taken place, and new things I declare; <u>before they spring into being I announce them to you</u>*** (Isaiah 42:9) .

- ***For I know** <u>the plans I have for you</u>**, declares the Lord...*** *(Jeremiah 29:11)* .

Thus, just as the growth and functioning of the human body and the bodies of animals are simply the execution of a pre-established program, i.e. the DNA contained within them, which enables them to manufacture their own molecules from the atoms resulting from digestion, so God's creation of the world was also the execution of a pre-established program: a series of orders given, written down and sealed, which made it possible to reconstitute the dust from the Lamb into the universe.

Revelation 5:1-2 : ***Then I saw in the right hand of Him who sat on the throne a scroll with writing on both sides and sealed with seven seals.***
I saw a mighty angel proclaiming in a loud voice, "Who is worthy to break the seals and open the scroll?"

There it is! **The scroll sealed with seven seals!** If Elohim had not revealed this to us, we would certainly know very little about the fact that the creation of the world was a sequence of programmed events—a pre-established plan unfolding step by step.

Creation was the execution of a series of commands—a sequence of words spoken in advance by Elohim, with the purpose of giving form to the dust that would come from the disintegrable part of the Lamb. It was therefore certainly not a series of spontaneous events that required Elohim's intervention at each new stage of creation.

In other words, Elohim already knew what the final universe would be before beginning its creation; the blueprint, the code of the universe, had already been designed through His spoken declarations, which had been written down, each one destined to be fulfilled in its appointed time.

Indeed, several other accounts in the Bible also reveal to us this way of doing things by Elohim: that of declaring things in advance and putting them in writing before they are carried out. This is what we see, for example, with Ezekiel the prophet, and also with John, the author of Revelation: everything they were to recount had already been prepared beforehand and written down by YHWH.

- ***And He said to me, "Son of man, eat what is before you, eat this scroll; then go and speak to the people of Israel** (Ezekiel 3:1) **."***

- ***I took the little scroll from the angel's hand and ate it... Then I was told, "You must prophesy again about many peoples, nations, languages and kings** (Revelation 10:10-11) **."***

Why would the scroll with seven seals seen by John likely be the DNA—the plan—of the universe?

The scroll with seven seals seen by John (Revelation 5) is most likely the blueprint of creation, the DNA of the universe, because we observe that starting from Revelation chapter 6 , the Lamb of God begins to open the seals of the scroll, and with each seal He opens, a series of catastrophes unfolds.

Catastrophes occur with each seal opening, until the opening of the sixth seal of the scroll, which results in a great upheaval of the cosmos, and also of the Earth's geology.

> *I watched as He opened the sixth seal. There was a great earthquake. The sun turned black like sackcloth made of goat hair, the whole moon turned blood red, and the stars in the sky fell to earth, as figs drop from a fig tree when shaken by a strong wind.*
> *The heaven receded like a scroll being rolled up, and every mountain and island was removed from its place* (Revelation 6:12-14) .

The opening of the seals thus creates an upheaval of established orders, a little as if a house constantly linked to its blueprint and standing only because its blueprint exists, were to see the blueprint gradually erased.

Yes, this is YHWH's way of doing things: He gives the plan of things in advance, before bringing them into existence by providing the Lamb, the One from whom every atom of the visible universe and every atom of the invisible universe derives its source. And this is exactly what He did before He began the destruction of the present universe and the creation of a new universe and a new heaven; He announced the creation of new things.

> ***See, the former things have taken place, and new things I declare; before they spring into being I announce them to you*** *(Isaiah 42:9)*.

Indeed, a large portion of the Old Testament (the first part of the Christian Bible, and also the Jewish Bible or Tanakh) is a proof of it: If, during the design of the blueprint for the current world, humanity had not yet been created, then for the destruction of this world and the creation of a new one, it pleased Elohim to reveal His new plan to humanity, which now existed.

Much of the Old Testament can be seen as an announcement of the coming of a new universe, alongside the renewal of human souls and bodies that will be brought into it—souls and bodies which the prophetic books often depict metaphorically as cities, towns, gardens, lands, mountains, or houses.

> ***This is what the Sovereign Lord says: On the day I cleanse you from all your sins, I will resettle your towns, and the ruins will be rebuilt.***
>
> ***The desolate land will be cultivated instead of lying desolate in the sight of all who pass through it .***
>
> ***They will say, "This land that was laid waste has become like the garden of Eden; the cities that were lying in ruins, desolate and destroyed, are now fortified and inhabited.***
>
> ***Then the nations around you that remain will know that I the Lord have rebuilt what was***

destroyed and have replanted what was desolate. I the Lord have spoken, and I will do it *(Ezekiel 36:33-36)* .

Yes, if Elohim planned in advance before the sacrifice of the Lamb in order to create a new world, then surely, for the creation of the present world, the Lamb was not sacrificed without a prior plan either.

And the existence of the blueprint for the first creation is most likely revealed to us through the account of the scroll with seven seals *(Revelation 5 and 6)*.

A few clarifications

If you have already understood that Elohim is one single Being in whom dwell the Father, the Lamb, and the Holy Spirit—also referred to as the Father and His Anointed Ones (*the two Sons of Oil*, according to Zechariah 4)—then you may already be starting to ask yourself certain questions:

- **Who among the Father, the Lamb, and the Holy Spirit spoke the words of the book sealed with seven seals?**
- **Who sealed the book, and with what?**

These are excellent questions!

Indeed, we must understand that, within Elohim, each of the Three has a distinct consciousness. Each One speaks, and they are intimately united with One Another—like an eternally watered land and the two olive trees growing from it, fulfilling a single will, all three sharing the same purpose. The eternally watered land is the Father, and His two olive trees are the Lamb and the Holy Spirit.

As mentioned earlier, before the Lamb went into the waters, a plan from the Father already existed—both for the Lamb Himself, and for the dust that would result from the dissolution of His disintegrable part in the waters.

So what was this plan?

It was a set of commands from the Father:

1. **Plan for the dust that came from the Lamb** = commands regarding the formation of the universe.

This plan was a sequence of commands spoken by the Father, written down and sealed—locked—by the blood (the life energy?) of the Lamb, in order to form the universe.

For, if God promised to seal the plan of the new world with the blood shed from the sacrifice of the Lamb, then it is certainly also by the blood of the Lamb that the commands defining the first world—the present world—were sealed.

> ***As for You also, because of the blood of your covenant, I will set your prisoner from the waterless pit*** *(Zachariah 9:11, NKJV)*.

2. **Plan for the Lamb Himself:** This plan was certainly a set of commands from the

Father, spoken over the Lamb (a part of the Lamb's genetic code?), declaring that the Lamb may lay down His life to the waters—and take it up again.

> ***Therefore My Father loves Me, because I lay down My life that I may take it again. None takes it from Me, but I lay it down of Myself. I have power to lay it down, and I have power to take it again. This command I have received from My Father*** *(John 10:17-18).*

Yes, it is certainly through the blood of the Lamb that the creative and foundational commands of the first world—the present world—were sealed by Elohim. And that is why it is only by the Lamb that we see these commands being progressively unsealed (the opening of the seals) to this day, for no other life is equal, none is greater, and no other blood is more precious.

> ***And I saw a mighty angel proclaiming in a loud voice, "Who is worthy to break the seals and open the scroll?"***

> *But no one in heaven or on earth or under the earth could open the scroll or even look inside it.*
>
> *I wept and wept because no one was found who was worthy to open the scroll or look inside.*
>
> *Then, one of the elders said to me, "Do not weep! See, the Lion of the tribe of Judah, the root of David, has triumphed. He is able to open the scroll and its seven seals* *(Revelation 5:2-5).*

Thus, as long as the relevant writings remain sealed, no other knowledge, no other power, no magic will be able to disturb the order of the universe in such a way as to cause galaxies to collapse for example. Because there are laws sealed by the life of the Lamb. And the life of the Lamb is the life of the Father, and the Father is greater than all.

> *For truly I tell you, until heaven and earth disappear, not the smallest letter, not the least stroke of a pen, will by any means disappear*

from the Law until everything is accomplished *(Matthew 5:18)*.

There it is! It has just been said: this is the very notion of the genetic code—the DNA—of the universe.
Just as the cells of our bodies contain a series of instructions telling them what to do at various moments of our lives, so too the entire universe is the materialization of a series of Father's commands given over the dust that came from the dissolution of the disintegrable part of the Lamb in the waters, and these commands were written down, then sealed—locked—by the blood (the life energy?) of the Lamb.

*[**Bonus**:*

Someone may say: If the whole universe with its immensity came from the Lamb, how immense (physically speaking) must the Lamb be!

It's a good concern! Indeed, the miracle of Jesus' multiplication of the loaves and fishes

*(**Matthew 14:14–21**) is a valuable teaching about a particular power of Elohim.*
In Elohim lies the power of multiplication: an atom can multiply infinitely to become hundreds of billions of galaxies, and hundreds of billions of galaxies can suddenly be reduced to the size of an atom.

So, as Jesus and his disciples watched those offering money in the temple ***(Luke 21:1–4)****, Jesus observed the widow and said that it was she who put in the most, for it's still not a matter of starting quantity as many may imagine.*
And Jesus said: ***It is easier for heaven and earth (that is, the entire universe) to disappear, than for the least stroke of a pen to drop out of the Law (Luke 16:17).*** *Yes—certainly because, among many other reasons, the laws that uphold heaven and earth come from YHWH, and therefore, a single stroke of a letter in the Law, though seemingly small, is greater and more powerful than what appears to us as*

infinitely vast (the whole universe).

That being said, at the foundation of the world—especially regarding the physical part of the universe—the Lamb could have cast Himself into the waters of death bearing the physical substance of a single atom, or even smaller than that, and then seen the dust resulting from His disintegration multiply and spread across astronomical distances, those of the universe, declared in advance by the Word (the orders, the plan) of the Father.

The smallest possible material particle was therefore enough to create the physical universe as we know it; what mattered most was simply that this particle was. For just like in the multiplication of the loaves and fishes, it's not about how many loaves and fishes there are, but about their availability and the power of the One who can multiply them at will.]

CHAPTER 7: AFTER THE WATERS

YHWH is truly Powerful! From destruction, from absolute chaos, from a massive black hole—in short, from death and utter despair—He brought forth worlds and their peoples, simply by speaking. Yes, at the Word of YHWH, things do not merely obey—they are executed and come into alignment; for His Word is a mighty current, before which the word *"resistance"* does not exist, and this current carries all things toward one single destination: what YHWH has spoken.

> ***For the Word of God is living and powerful, and <u>sharper than any two-edged sword</u>...*** *(Hebrews 4:12)*.

Much earlier, drawing on various biblical references, we saw that during the creation of the world, Elohim cast His Thinking Arm (the Lamb of God, Yeshua) into the waters of death contained within the abyss and covered by darkness, and that the part of Him reducible to dust disintegrated in those waters, with the dust produced by that disintegration serving to form the universe.

It is therefore at this stage of absolute chaos and destruction that the account reported in the first chapter of the book of Genesis begins.

> ***In the beginning God created the heavens and the earth. The earth was without form, and void; and darkness was on the face of the deep. And the Spirit of God was hovering over the face of the waters*** (*Genesis 1:1-2*) .

Thus, at this stage, concerning both the immaterial and physical universe, we had dust (atoms?) suspended in the waters of death, which themselves were covered with clouds (Genesis 1:1-2) and darkness.

DIAGRAM 7: THE WATERS WRAPPED IN CLOUD AND DARKNESS

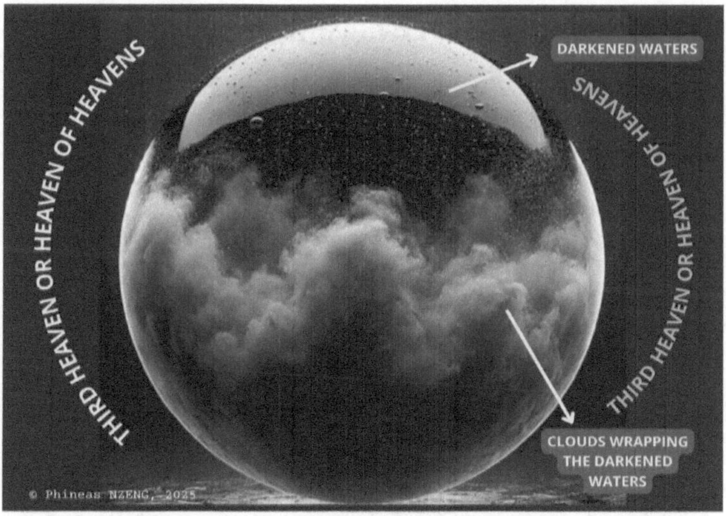

Illustration of the First Earth (reproduction of Diagram 6): *We see waters covered with darkness and clouds. Within the waters is suspended dust, but it cannot be seen here because of the darkness.*

This was ***the first earth****, the one mentioned in* ***Genesis 1:1*** *("In the beginning God created the heavens and the earth"), which was destined to become the second heaven and the universe.*

This first earth was actually a massive black hole.

Yes, it is from this first earth that Elohim formed **the universe** (the planets and their skies, which make up the first heaven) as well as **the**

second heaven, in the later stages of creation. The third heaven, however, was not formed within this structure, but beyond it—beyond the clouds and the darkness—and it is most likely this third heaven that is referred to in Genesis 1:1 ("In the beginning, God created **the heavens** and the earth").

Earlier, we also saw that, in terms of the material and physical universe, the waters of death are most likely the substance that makes up the black holes of the universe. And as for the immaterial world (the realm of the soul), the waters of death are the Sheol — the place where souls not admitted to God's rest are reduced to dust and chaos (Psalm 22:15 & Psalm 69:13–16).

The Next Stages of Creation

As previously seen, Elohim — the Author, Planner, and Creator of the universe — had certainly planned everything in advance and put it into writing even before creation itself.

The dust that came from the Lamb, along with the waters that contained it, were therefore programmed to follow the commands and instructions spoken and written beforehand by the Father, and sealed by the blood (the life?) of the Lamb.

[Indeed, most Bible translators have used the present tense to translate Elohim's commands, highlighting the simultaneity between the orders given and their execution — and this remains accurate. For example: ***"God said, 'Let there be light'*** *(the declaration is in the present tense),* ***and there was light*** *(Genesis 1:3)."*

However, since another wording is also possible, some authors used the future tense to translate Elohim's commands, bringing out, therefore, the precedence of Elohim's commands over the elements created by them. A rare Bible translation that reflects this is Chouraqui's Bible, which is a French version in which the translator renders the commands in the future tense. Here's what **Genesis 3:1**, *for example, would look like, if translated from Hebrew into English following Chouraqui's pattern:*

Elohim said: **"*A light will be*"** *(the command is in the future),* ***and here is a light*** *(Genesis 3:1).*

This highlights the fact that the command for the light to appear had been given well before its execution.

Shall we then say that YHWH cannot also give commands that are fulfilled instantly? Far from it—for it is also written that YHWH does whatever He pleases, and nothing is impossible for Him.

Whatever the Lord pleases He does, in heaven and in earth, in the seas and in all deep places *(Psalm 135:6)* .

In the case of creation, it certainly pleased YHWH to first establish a plan.]

Thus, under the command of YHWH, the following were created:

- **The light**

By the command of YHWH, within the mixture of dust and the waters of death, light appeared.

But light was still mingled with darkness, so YHWH made it so that, in the course of time, wherever a light would be kindled, the darkness would flee: He therefore separated the light from the darkness, and this marked the beginning of Era 1 of creation — the first day of creation.

> ***And God said, "Let there be light", and there was light.***
> ***God saw that the light was good, and He separated the light from darkness.***
> ***God called the light "day," and the darkness He called "night." And there was evening and there was morning—the first day*** (Genesis 1:3-5) .

"There was an evening and there was a morning" is an expression that can also signify: **"A time of implied preparation came to an end (there**

was an evening), and new things came into existence (and there was a morning).”

Thus, throughout the first chapter of Genesis, the phrase “a morning” refers to each major new event that eventually came into being (a morning) after a quiet time of preparation came to its end (an evening).

- **The Second Heaven** (Genesis 1:6–8)

By the command of YHWH, within the mixture of waters and dust, a space—an expanse, a vault—was formed, separating the waters into those above and those below. And since we are most certainly dealing with a sphere, this simply means that a hollow pocket was created in the middle of the sphere, revealing waters both external and internal to this newly formed space.

DIAGRAM 8: CREATION OF THE SECOND HEAVEN

Below is half of a sphere cut through its middle in a sagittal way.

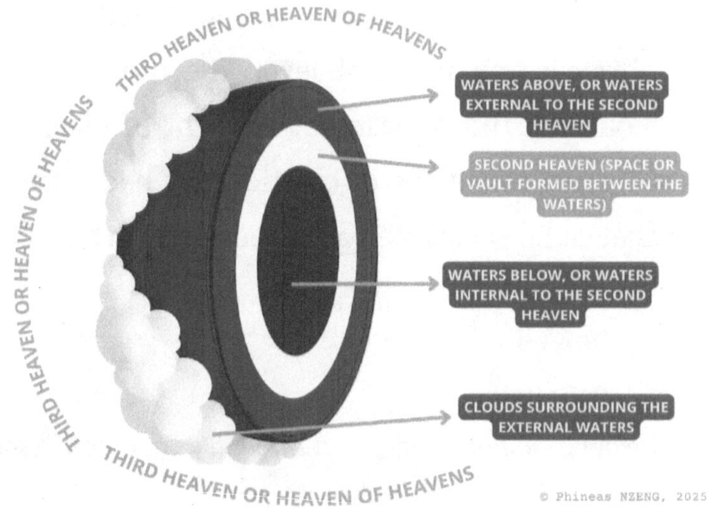

***Note:** Let us remember that at this stage, the waters are darkened and contain dust—the dust that came from the Lamb.*

YHWH called this newly formed space "heaven." We must keep in mind that this heaven is different from the one mentioned at the very beginning of Genesis 1 ("In the beginning, Elohim created **the heaven** and the earth"); the heaven first mentioned was certainly the Heaven of heavens, or the third heaven, where YHWH has His throne. This newly formed space was therefore the second heaven.

Now remember what we said earlier: on the physical and material level, the waters of death would be the constituent element of black holes. Thus, at this stage of creation, We had a structure that can be described as in the diagram below:

DIAGRAM 9: OVERVIEW OF THE WORLD ON THE SECOND DAY

Below is half of a sphere cut through its middle in a sagittal way.

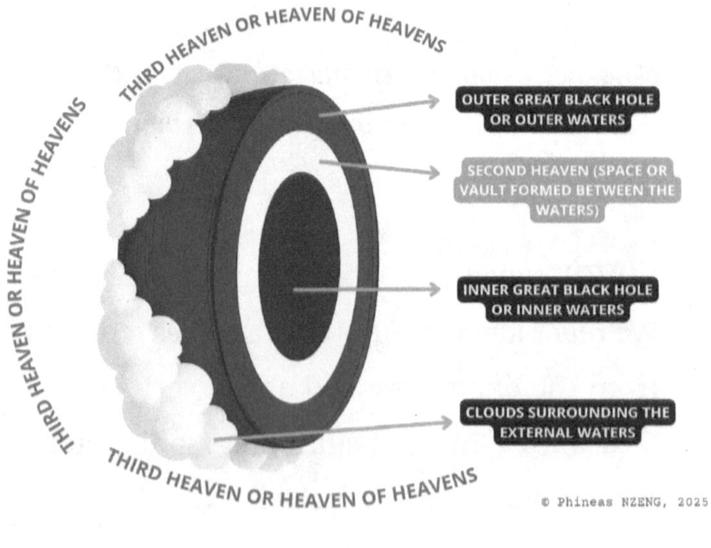

Note: *Let us remember that at this stage, the waters are darkened and contain dust—the dust that came from the Lamb.*

> ***And God said, "Let there be a vault between the waters to separate water from water."***
> ***So God made the vault and separated the water under the vault from the water above it. And it was so.***
> ***God called the vault "sky*** *('heaven' in many other versions)."* ***And there was evening, and there was morning—second day*** *(Genesis 1:6-8).*

- **The Universe and Its black holes** (Genesis 1:9–10)

At the command of YHWH, the inner waters—or those beneath the second heaven—separated from the dust; the inner waters gathered together and allowed the dry land to appear.

Here, we should not understand 'dry' as 'that which is not wet' in the literal sense. Rather, we should understand it as any part under the second heaven that no longer contains the waters of death. This dry land was therefore dust now suspended in the void, no longer within the waters of death. In the physical and material sense, this 'dry land' referred to 'any part that was no longer a black hole under the second heaven.

YHWH called the dry land 'Earth,' and He called the gathered waters—separated from the dust—'Seas.'

Have you further discovered what this 'dry land' called 'Earth' by YHWH truly meant?

This 'dry land' was the future stars and planets, all the non–black hole regions of the future galaxies, as well as all intergalactic, interstellar, and interplanetary space—in short, everything that is not a black hole.
And the Seas, in this context **(God called the gathering of the waters 'Seas'...)**, are the black holes of the universe.

DIAGRAM 10: THE UNIVERSE AND ITS BLACK HOLES

Below is half of a sphere cut through its middle in a sagittal way.

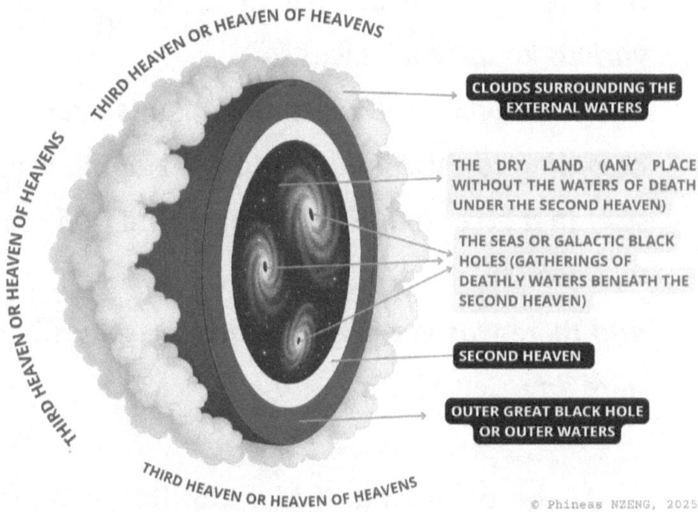

There it is! We have just seen the **second 'earth'** mentioned in **Genesis chapter 1**. If the first 'earth' referred to the totality of the waters of death suspended in the deep, containing dust and covered with darkness and clouds (see Diagram 7 above), then this second 'earth' refers to the dust and space completely separated from the waters of death.

- probably several habitable planets

Then God said, "Let the land produce vegetation: seed-bearing plants and trees on the land that bear fruit with seed in it, according to their various kinds." And it was so.
The land produced vegetation: plants bearing seed according to their kinds and trees bearing fruit with seed in it according to their kinds. And God so that it was good.
And there was evening and there was morning—the third day (Genesis 1:11-13).

At the command of Elohim, the earth brought forth vegetation; greenery appeared.

So, which earth was being referred to here, at a time when the stars, including the sun, had not yet been formed? Was it the planet Earth that our feet tread upon today? Several other planets as well, grouped together or scattered? Or the entire universe?

The Bible does not give us these details.

However, we are certain of three things:

- According to our human understanding, dust must be gathered into a compact mass for vegetation to grow on it. Therefore, the land referred to here must have been, from a human perspective, one or several rocky planets, grouped together or scattered, which, at that time, had already been formed and bore vegetation.

  However, we cannot state this with certainty, for nothing is impossible for YHWH, and even on dust in suspension He can cause vegetation to grow.

- At that time, the stars—including the sun—had not yet been formed, but the light already existed to sustain vegetation (from a human perspective). Thus, just like the first two instances of 'earth' mentioned in Genesis chapter 1 , this third 'earth' mentioned was necessarily not only the planet Earth as we know it today; several other planets bearing vegetation may also have been included in this term “earth that brought forth vegetation.”

- I may not be the only one to think that the human being originally created by YHWH was very likely an interplanetary—perhaps even intergalactic—being, and that it was because of his fall, because of the original sin, that he became confined to the Earth (the planet Earth), slave to gravity and the elements. The entire universe was therefore most likely his home—several other planets bearing vegetation may have been reserved for him. For somewhere, speaking of how YHWH made man, it is written:

  What is mankind that you are mindful of them, human beings that you care for them?

  You have made them a little lower than God and crowned them with glory and honor (Psalm 8:4-5).

The vegetation mentioned in the section of Genesis 1:11–13 may therefore have concerned multiple planets across various galaxies—perhaps even all the planets in the universe—the rocky as well as the gaseous ones.

- **The Luminous Celestial Bodies (Genesis 1:14–19)**

At the command of YHWH, the sun, the moon, and the stars were created on the fourth day. These created elements, listed from verse 14 of Genesis 1 onward, refer specifically to planet Earth—our planet, the one upon which the soles of our feet walk. For in this passage, the author describes two great lights: one to rule the day—clearly referring to the sun—and the other to rule the night—clearly referring to the moon.

> ***Then God said, "Let there be lights in the firmament of heavens to divide the day from the night; and let them be for signs and seasons, and for days and years; and let them be for lights in the firmament of the heavens to give light on the earth;" and it was so.***

Then God made two great lights: the greater light to rule the day, and the lesser light to rule the night. He made the stars also.
God set them in the firmament of the heavens to give light on the earth, and to rule over the day and over the night, and to divide the light from darkness. And God saw that it was good.
So the evening and the morning were the fourth day (Genesis 1:14-19, NKJV).

Did you notice? The heaven spoken of here—the one where the two great lights and the stars were created—is **the first heaven,** our sky, the one above our heads, the sky that fills with stars on clear nights.

This first heaven is, in fact, the entire portion of the universe that surrounds the Earth—and of which the Earth itself is a part. It is the region that contains the galaxies, and it is the part of the universe accessible to our probes, telescopes, and other tools of space exploration.

Beyond the first heaven—that is, beyond our observable universe—lies a vast void whose nature we do not know: **the second heaven**. It is this great expanse that YHWH created to separate the waters from the waters—do you remember? (Genesis 1:6–8) .

DIAGRAM 11: SUMMARY OF THE THREE HEAVENS

Below is half of a sphere cut through its middle in a sagittal way.

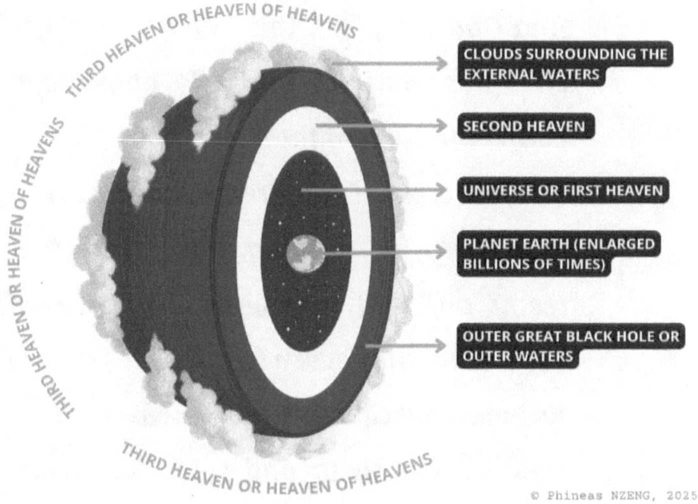

- **Fish, Birds, and Land Animals** (Genesis 1:20–25)

Now remember, we said earlier that the plan—the sequence of commands to be followed—had certainly already been designed by the Father before the Lamb was sacrificed to the waters, so that His breakable part would be reduced to dust. It was therefore upon that dust (atoms?) from the Lamb that the Father's commands continued to unfold here, bringing into existence the fish, the birds, and the land animals.

> ***And God said, "Let the water teem with living creatures, and let birds fly above the earth across the vault of the sky."***
>
> ***So God created the great creatures of the sea and every living thing with which the water teems, and that moves about in it, according to their kind, and every winged bird according to its kind. And God saw that it was good.***
>
> ***God blessed them and said, "Be fruitful and increase in number and fill the water in the seas, and let the birds increase on the earth."***

And there was evening and there was morning—the fifth day.
And God said, "Let the land produce living creatures according to their kinds: the livestock, the creatures that move along the ground, and the wild animals, each according to its kind." And it was so.
God made the wild animals according to their kinds, the livestock according to their kinds, and all the creatures that move along the ground according to their kinds (Genesis 1:20-25).

The earth, the sky, and the water mentioned here all relate to our planet—the ones we are all familiar with.

And let us notice this: to create the animals, and even the plants, YHWH used only the land—that is, only the dust that came from the Lamb.

And God said, "Let the land produce living creatures according to their kinds: the livestock, the creatures that move along the

> ***ground, and the wild animals, each according to its kind." And it was so*** (Genesis 1:20-24) .

Indeed, the Lamb is the Logos — the Word that possesses intelligence — and what YHWH produced from His dust (stars, planets, plants, animals, humans) is thinking, conscious, or possesses instinct, somewhat like a lizard's severed tail continuing to move. However, in the immaterial dimension, only the soul (the immaterial body) of the human possesses the divine element that is in the Lamb. Thus, in the waters of death, an animal or plant soul is certainly destined to completely disintegrate and return entirely to dust, but in the human, something remains forever.

So, when YHWH shows Ezekiel the corpses of humanity in the valley (Ezekiel 37) , He shows him bones—something that remained from the souls; and Isaiah, speaking of the human souls that will be sent to hell, writes:

> ***Therefore, as the fire devours the stubble, and the flame consumes the chaff, so their root will be as rottenness*** *(their source part will just rot, but not disintegrate),* ***and their blossom will ascend like dust*** *(the rest of their being will fall apart);* ***because they have rejected the law of the Lord of hosts, and despised the law of the Holy one of Israel*** *(Isaiah 5:24, NKJV)*.

Yes, the flower of the human soul is reduced to dust, but its root—that is, the bones of the immaterial body, the divine part of the soul (the part meant to keep the soul attached to the Lamb?)—is not reduced to dust; it only decays, yet remains. It is somewhat like a living skeleton that retains full consciousness and can feel the pain of its flesh and skin being constantly lost.

However, **the Good News** is that when Jesus Christ went into the waters a second time, the dust of His Soul restored to the flesh on our bones (the flower of our root) the freshness lost because of the disobedience that had separated us from the

Source (the Father) and from His Trunk (Jesus Christ, the Lamb).

> ***I am the Living Bread that came down from heaven. Whoever eats this bread will live forever. This bread is my flesh*** *(flesh of the soul above all)*, ***which I will give for the life of the world*** *(John 6:51)*.

From now on, all those who turn away forever from their conscious sins (repentance), ask God for forgiveness, and confess with their mouth and with their spirit the Lord Jesus over their being, can see their flower regain its freshness and no longer be dry and to be burned forever.

But it is a matter of consistency: in addition to repenting and confessing Jesus, the root of our flower must remain connected to the Source through daily and regular reading of the Bible, daily and regular prayer to the Father in the Name of Jesus, constant obedience to God, as well as daily and regular praise and worship to God. Whoever doesn't do all this, at the resurrection of the dead, will be like the virgins (people who even received Jesus) who had not kept oil in their lamps before falling asleep, this means before dying (Matthew 25).

Thus were created the three heavens, the earth, the plants, and the animals—and through this, the ground for welcoming mankind had been prepared.

CHAPTER 8: THE HUMAN

Miniature Versions of the Lamb...

Earlier, we spoke about the Lamb of YHWH—the One from whom our souls and everything physical in the universe originated: every grain of dust in the galaxies, every particle of the tools and utensils in our homes, every atom of our bodies—in short, everything that can exist physically in the trillions of galaxies that make up the universe (planet Earth and the first heaven) came from the Lamb.

And as we said earlier, the creation story reported in Genesis 1 actually begins with a land already completely destroyed ("the earth was tohu wa-bohu..."); this earth was the perishable part of the Lamb, which had become dust (atoms?) in the waters of death.

And from this dust—this dissolved earth—planet Earth and the first heaven were created, but not the second heaven, which is a space created between the waters **(see Diagram 8 in the previous chapter)**. As for the third heaven, the Bible gives us no further detail beyond the fact that it is mentioned.

Was there still some of the Lamb's dust left in the waters beyond the second heaven after creation? Perhaps so—but the Bible neither confirms nor denies it.

> *[Bonus: Are the Waters of Death stronger than Yeshua (the Lamb)? Do they have power over Him?*
> ***Absolutely never!*** *Indeed, just as the Father has the power to dry up the waters, so does the Son (the Lamb, Yeshua), the Holy Spirit, and all those who bear the same Name as the Father through the Son and the Holy Spirit—that is, Christians and the angels of YHWH.*
>
> *Thus, in the Name of Jesus Christ, we dry up the streams, rivers, and torrents that carry souls*

into the great sea—that is, the Sheol, the realm of the dead. These streams, rivers, and torrents are: sickness, accidents, witchcraft, curses, poisonings, and the like.

> ***Who (YHWH) says to the deep: "Be dry, and I will dry up your rivers (Isaiah 44:27) ."***

Both for the creation of the world and for its salvation and re-creation, the Lamb—the Thinking Arm of YHWH—was torn apart by the waters, simply because He chose to let it happen.

He chose to remain silent, for such was the mission—the mission of YHWH's infinite Love. Without this mission having been fulfilled the first time, you and I would never know what it feels like to hold a beautiful baby in our arms, or to enjoy a good night's sleep after a long day—because the world, and all its cherished little joys, would not exist.

And without this mission being fulfilled a second time, hell would have been the final destination of every human soul.
YHWH's plan, therefore, was that His Thinking Arm remained silent before the waters.

> ***He was oppressed and afflicted, yet He did not open His mouth; He was led like a lamb to the slaughter, and as a ship before its shearers is silent, so He did not open His mouth** (Isaiah 53:7).]*

As previously mentioned, the Lamb has a part that cannot be disintegrated by anything **(Isaiah 53:7)** . And in addition to His indestructible part, the Lamb of God—Yeshua ha Mashiach—has an earthly and perishable part, both material and immaterial, which the material waters (the substance of black holes?) and the immaterial waters of death (the waters of Sheol) can reduce to dust.

So, you know what? We human beings and the Lamb are of the same kind.

Just as the Lamb possessed, from eternity, a physical substance of a nature known to YHWH Himself—and whose disintegration gave rise to the entire physical universe—in the same way, human beings also possess a physical substance: the human body, which the Lamb Himself wore when He came to dwell among us.

And from the dissolution of the disintegrable immaterial part of the Lamb (from the earth of the lamb's Soul) came the disintegrable part of our souls, as well as the entire immaterial universe.

Our souls also have a part that the waters of death can never destroy, a part that most certainly grew like branches from the Lamb at the moment when YHWH created human souls—a moment known to Him alone.

> ***I'm the Vine; you are the branches. If you remain in Me and I in You, you will bear much fruit. Apart from Me you can do nothing*** *(John 15:5)*.

Indeed, your soul—the "you" of your dreams and visions—is also a body; you therefore have a body within another body, and the Bible speaks at length and in depth about what our souls are, and about the double beings that we truly are.

> [*70 Centuries to Go for the Universe: The Reparation* is a book project by the same author (Phineas NZENG), which explores in depth the subject of what the soul is, and I highly recommend it to you as soon as it is released, if it has not already been.]

Thus stated, we will now briefly examine the parts of the human being—especially the immaterial ones—according to what the Bible teaches us.

The Parts of the Human Being...

- **The Body**

We don't even need to explain that! Right? So, let's quickly move on...

- **The Soul**

Earlier, we mentioned that most of the elements described in **Genesis 1** should be considered as belonging both to the immaterial world and the material world, based on the double nature of the Word of God as stated in **Hebrews 4:12**.

Thus, with regard to dust and water, for example, we have seen that they are elements that are both material and immaterial. And if it is written that man was made from the earth **(Genesis 2:7)**, then the human being must be understood as having both a body made of material earth and a soul made of immaterial earth.

> *[**Bonus:** Unlike the Lamb, the human being who came into the world had most likely never known the waters of death; and perhaps that is why he did not take very seriously the warning of death tied to disobedience **("On the day you eat from the tree of the knowledge of good and evil, you will surely die..." — Genesis 2:17**).]*

The soul is therefore an entity made of dust, and most importantly, it is not something shapeless; just like our body, it also has parts—and those parts have specific functions. And this is exactly what the Bible teaches us.

Indeed, several terms are used in the Bible to refer to all or parts of our inner being. For example: mountain, city, house, garden or wilderness, vineyard, tree, flower, tabernacle, temple—terms that are very often used in the books of the prophets.

Of all these terms, we will choose just one and give a very brief description of it; this will help us understand what YHWH says we are on the inside—what the human being looks like when seen by a spirit. Below, the chosen term:

THE TABERNACLE (Exodus 25–40)

As YHWH began to reveal Himself more and more to humanity through the people of Israel, it pleased Him to reveal to human beings what each person truly is deep within—He revealed to them the anatomy of our immaterial being.

YHWH gave Moses the command to build a tabernacle—a portable and movable place of worship—a physical reflection of YHWH's true temple (the human being), where the people, with the help of the priests, were to worship YHWH while they were in the wilderness, journeying toward the Promised Land.

DIAGRAM 12: OVERVIEW OF THE TABERNACLE

The tabernacle desired by God was made up of two main components: it was **a tent** and its surrounding enclosure, called the "**courtyard.**"

- **The tent** was like a small house **with two rooms:**

1. **The Holy Place** – the first room, immediately after the entrance of the tent; it contained three essential objects: a **table holding bread**, an **altar for burning incense**, with a horn on each of its four corners, and a **lampstand with seven lamps** (the menorah), the only light in the room.

DIAGRAM 13: THE HOLY PLACE (FIRST PART OF THE TENT)

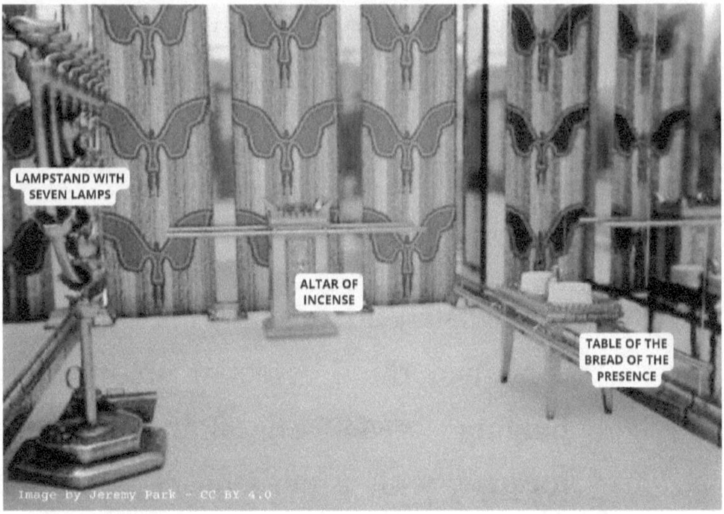

Illustration of the Holy Place as it was seen from the entrance of the tent. The curtain, embroidered with cherubim at the back of the room, marked the separation between the Holy Place and the Most Holy Place. Here, the objects in the room are reflected on the golden surfaces of the left and right sides of the room.

2. **The Most Holy Place (Holy of Holies)** – the second room of the tent. It was separated from the Holy Place by a large curtain embroidered with cherubim. Inside this room was **a single essential object: the Ark**

of the Covenant.

And inside the Ark of the Covenant were **the tablets of the Ten Commandments,** alongside **the manna** and **Aaron's dry staff, on which flowers and fruits had miraculously blossomed.**

DIAGRAM 14: THE MOST HOLY PLACE OR HOLY OF HOLIES (SECOND PART OF THE TENT)

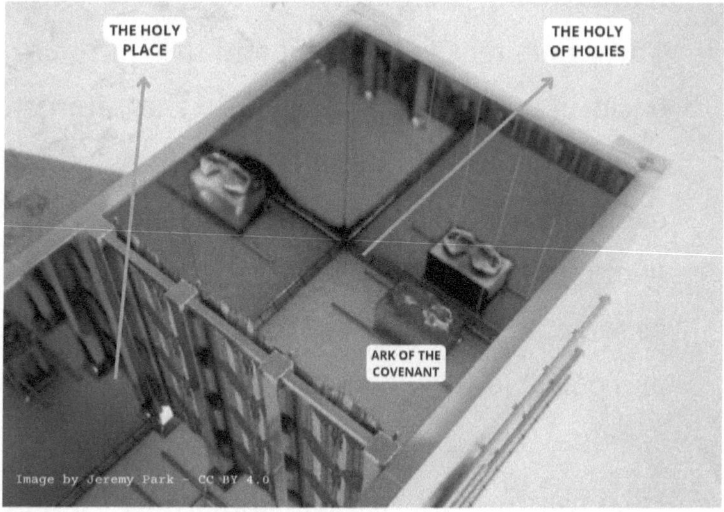

- **The courtyard** is the interior of the enclosure of the tent **(see diagram 12 above)**, and it is likewise full of teachings, which we address in the work *70 Centuries to Go for the Universe: The Reparation.*

__Explanations__

Indeed, the entire tabernacle represents what the human being is inwardly:

The whole tent is the soul, and the courtyard (the enclosure of the tent) is the immaterial representation of our body as well as of various elements related to the field of influence of our soul (work, marriage, health, possessions, etc.). And just as we are mobile, so the tabernacle was movable and portable.

Now, let us briefly say something about the objects of the Holy Place:

- *The altar of incense*

DIAGRAM 15: THE ALTAR OF INCENSE

The altar of incense, located in the first part of the tent, had four horns, one at each corner, facing outward. These horns represent the emotional and affective struggles, the inner dispersions, that every human being must neutralize in order to live a fulfilled life.

These horns certainly did not exist before humanity's sin, for they are surely the 'thorns of the crown' that Jesus Christ wore when He died (Matthew 27:29).

And it must be said that only confessing, in the heart, the blood of Jesus Christ upon these horns can truly neutralize them.. This is why YHWH commanded that the blood of animals be applied to the horns of the altar of incense, as a foreshadowing of the perfect sacrifice to come: that of Yeshoua ha Mashiah.

Yes, Yeshua wore a crown of thorns, for He took upon Himself our emotional and affective struggles (anxieties, depressions, worries of all kinds, addictions, etc.) in order to give us His peace through His life, through His blood. We must therefore confess in faith the blood of Jesus—which is the life of Jesus—against our worries of every kind.

> ***... if you confess with your mouth the Lord Jesus and believe in your heart that God has raised Him from the dead, you will be saved.***
> ***For with the heart one believes unto righteousness, and with the mouth confession is made unto salvation*** (Romans 10:9-10, NKJV).

Also, the incense burned on the altar of incense represents the fruit of thoughts turned toward YHWH and focused on Him and His Word: praise and worship, prayers and a life of integrity, proclamation of the Word of faith—the Word of God. All of this, coupled with the confession of the blood of the Lamb, should be done every day, in an attitude of renunciation of all conscious sin, in order to eliminate all incense altars that may form in your tent, as it is written:

> ***They triumphed over him by the blood of the Lamb and by the word of their testimony ; they did not love their lives so much as to shrink from death*** *(Revelation 12:11).*

Indeed, an incense altar is, in reality, a demon's head in a tabernacle; it is equipped with horns to frighten and scatter the human spirit and mind, and it is filled with fire—that is, the waters of death (the ideas of the realm of the dead)—to cause harm. Therefore, neutralize its horns through the confession of the blood of Jesus Christ, dry up its waters by the word of your testimony *(burn incense, as defined in the previous paragraph)*, which is a fire to its waters of death, and a water to its fire. And be ready to abandon forever (to kill) anything that makes you blameworthy.

This is how one crushes the horned head of the serpent!

> ***I will put enmity between you and the woman, and between your offspring and hers; he will crush your head, and you will strike his heel*** *(Genesis 3:15)*.

> [Do you wish to know what it means to bruise the heel? Then I invite you to meet me again in *70 Centuries to Go for the Universe: The Reparation*.]

Thus, the less you pray, the less you praise God through Jesus Christ, and the less you think about His sacrifice for you in order to confess it, the more your peace will be of very poor quality, and you will turn to worldly means (drugs, debauchery) to find a false tranquility of the soul.

God had commanded that the sacrifices offered in the tent be perpetual. Therefore, offer sacrifice every day of your life: say "Jesus" every day, and confess His blood over every aspect of your life; worship God; pray and praise Him in all circumstances. Above all, always do all this in sanctification and with a grateful heart, for the blood of Jesus has already been shed and remains forever available.

Those who sacrifice thank offerings honor Me...
(Psalm 50:23)

- ***The seven-lamp lampstand***

DIAGRAM 16: THE SEVEN-LAMP LAMPSTAND

Seven-branched lampstand, with its reflection seen on the golden walls of the tent.

The seven-branched lampstand is truly a lampstand bearing seven lamps, which every human soul carries within itself.

Indeed, the seven-branched lampstand was made of a central shaft, to which three branches were attached on each side: three branches on the left and three branches on the right.

Along the length of each of the three branches connected to the central shaft were depicted three fruits—almonds—each one topped with a flower.

There were also almonds with blossoms on the central shaft, numbering four, unlike the other branches, which had three (Exodus 25:31–40).

Unlike the other objects of the tent, and even the walls of the tent, which were made of acacia wood and overlaid with gold, the lampstand was entirely of gold.

Indeed, gold symbolizes the spirit, and the representation of everything that is wholly spirit was made of pure gold in the tabernacle, as in the case of the two cherubim depicted on the cover of the Ark of the Covenant.

Each of the seven branches of the lampstand was topped with a lamp, and the lamps were fueled with olive oil and kept burning from evening until morning by Aaron and his sons, perpetually (Exodus 27:20–21).

Whoever speaks of a lamp speaks of light, and whoever speaks of light speaks of discernment, knowledge, and visualization.

The six dimensions...

As can be understood on the basis of what has been said in the previous chapters, the world created by YHWH comprises a total of six dimensions, which are:

- **Three material dimensions:** the physical universe (earth + first physical heaven) + the second and third physical heavens. In the physical world, distances are so immense that they exceed the comprehension of present-day humans.

- **Three immaterial dimensions:** the immaterial universe (earth + first immaterial heaven) + the second and third immaterial heavens. There is no distance between immaterial entities.

The seventh dimension...

Beyond the earth and the physical and immaterial heavens stands YHWH, Creator of the worlds in their different dimensions.

YHWH is in the seventh dimension, the one from which He created the worlds, and from which He constantly sees them, as a painter sees his work of art.

Being in the seventh dimension, YHWH is constantly aware of everything: no ant walks upon the earth without His knowing it; no microbe exists in a living organism without His knowing it; no star goes out in a galaxy without His knowing it; no soul decays in the waters of Sheol without His knowing it. He knows each of His servants who rest today while awaiting the resurrection, and as for you and me, He knows perfectly every detail of our lives and loves us above all.

> ***Are not two sparrows sold for a penny? Yet not one of them will fall to the ground outside of your Father's knowledge.***

And even the very hairs of your head are all numbered (Matthew 10:29-30).

Nothing perfect can exist unless it has first been examined and approved from the seventh dimension, the one from which everything is seen and known at the same time; for from the other dimensions one can have only a partial vision, knowledge, and understanding of things, forever.

It is YHWH alone (that is, the Father, the Lamb, and the Holy Spirit) who is the Master of the seventh dimension, and everything He does or approves is declared good and perfect (Genesis 1:31), because His plans and His actions create no imbalance in any of the heavens or any of the earths. This is why all the plans of YHWH are Sabbaths; all the plans of YHWH are plans of dimension seven: plans that are truly complete, requiring no further adjustment, for YHWH cannot make an error.

All the plans of YHWH are Sabbaths because, once a plan has passed the examination of the seventh dimension, it is declared finished, and YHWH can rest from His work.

> ***By the seventh day, God had finished the work He had been doing; so on the seventh day, he rested from all his work*** (Genesis 2:2).

There is therefore a Sabbath in everything that YHWH designs, plans, and gives: a marriage, a job, a child, a calling, any kind of possession, and whoever interferes with a plan of YHWH in the tabernacle of another or in his own tabernacle through magic, sorcery, theft, jealousy, and the like, is a violator of the Sabbath of YHWH, and will not go unpunished.

> ***But if you do not obey me to keep the sabbath day holy by not carrying any load when you come through the gates of Jerusalem on the sabbath day, then I will kindle an unquenchable fire in the gates of Jerusalem that will consume the fortresses*** (Jeremiah 17:27).

The Jerusalem spoken of here is first and foremost the human being; it is the tabernacle.

Thus, all the commandments of YHWH governing relationships between human beings converge toward a single commandment: “Do not disturb what YHWH has finished doing! Respect the Sabbath day!” This may be why this commandment is the seventh when one counts the Ten Commandments starting from the last (Exodus 20:1–17).

Let us return to the seven-branched lampstand...
In creating human beings, YHWH endowed each of them with a lampstand in his tent (in his soul), a lampstand with seven lamps.

The number of lamps that are lit can signify the level of knowledge, discernment, or understanding of things, the number of realities perceived, and the degree of awareness one has of the worlds.

This lampstand is most certainly the lamp carried by each of the ten virgins in the parable of the ten virgins (Matthew 25:1–13); it is also the lampstand that the prophet Zechariah sees in the vision recorded in Zechariah 4.

And according to Zechariah chapter 4, it is by the Oil coming from the two Olive Trees that stand before the Lord of all the earth that the lamps of the lampstand are lit. The two Olive Trees are Jesus Christ and the Holy Spirit, and the Lord of all the earth is the Father.

Thus, **only those who have embodied Jesus Christ and who live in communion with the Holy Spirit** can see all seven lamps of their lampstand lit; for according to Zechariah chapter 4, it is through His two Olive Trees (Jesus Christ and the Holy Spirit) that the Father channels His Oil toward creation. And because the Father is in the seventh dimension, all discernment, all illumination, and all contemplation coming from Him belong to the seventh dimension and remain perfect and non-harmful for eternity, for they lead to no imbalance within the worlds.

> ***The blessing of the Lord makes one rich, and He adds no sorrow with it*** *(Proverbs 10:22).*

It is by believing in the sacrifice of Jesus—that is, in His body and His soul given over to death to heal our diseases and restore our souls—while confessing this for ourselves, while believing that YHWH raised Him from the dead, and while inviting Him into our hearts, that one comes to embody Jesus.

And those who have embodied Jesus Christ, the Father gives them His Holy Spirit, so that their lamps may be lit by the Oil received both from Jesus and from the Holy Spirit within them (Zechariah 4).

And for these lamps to remain lit, so that the human being may live eternally, it is necessary: every day, and without ceasing, to pray to the Father in the Name of Jesus Christ; to read the Bible regularly; to praise and worship YHWH regularly through songs and in prayer; and to obey YHWH always.

Whoever maintains his lamp in this way until old age and death will fall asleep in death with Oil in his lamp, and will rise at the resurrection of the dead with a lamp full of oil, like the five wise virgins in the parable of the ten virgins (Matthew 25:1–13).

The trunks, their branches, and the Oil...

To explain things in a bit more detail: Zechariah chapter 4 teaches us that there is the Father, who is the Planter of the two Olive Trees, whose Oil—otherwise called gold—feeds the seven-branched lampstand.

If the two Olive Trees are trees, they are not trunks alone: they have branches. One of the two Olive Trees is Jesus Christ, and His branches are human beings (John 15), and the other Olive Tree is the Holy Spirit, and its branches are the angels, for the angels are of the same nature as the Holy Spirit: angels are winds (Psalm 104:4).

In reality, it is through the branches, which are the final actors, that the light of YHWH (the seven lamps) is meant to be applied to the six worlds to govern them.

And it takes a joint action of the two branches through their descendants (angels and humans), all united by the perfection of the Father (the central shaft), for the light of the seven lamps to be lit.

Put differently: it takes the portions of Oil from the two branches through their descendants—namely, three portions coming from humans and three portions coming from angels—the six portions united by the portion of Oil that only the Father, the Son, and the Holy Spirit can provide (the seventh portion of Oil), for the seven lamps to be lit and for the world to remain perfect for eternity; the six dimensions of the world being governed by the light of the seven lamps of YHWH.

All the imperfection and suffering of the world exist solely because branches independent of the two trunks—that is, humans who have not embodied Jesus Christ (humans not attached to Jesus, humans who do not have the nature of Jesus, humans not born again) together with angels not attached to the Holy Spirit (demons, gods, idols, laws, ancestors, spirits)—take control and assume the administration of the world. Thus, the six is administered by the light of six lamps, resulting in chaos.

[*Bonus*:

1. *The human being, as a branch of Jesus Christ, was therefore never created to be the sole inhabitant of the tabernacle that he is; his spirit (the gold covering the walls and the objects) was created to cohabit there with the angels, so that, for each of his actions, the light of his three lamps may be added to the light of the three lamps of the angels, the two lights cemented by the light of the Father, so that the world may be governed by the light of YHWH, the light of the seven lamps.*

   Thus, across the entire ceiling of the Tabernacle tent (not represented in this work), cherubim were embroidered **(Exodus 26:1)**, *that is, angels—branches of the Holy Spirit and messengers of YHWH—who descend toward the lower parts of our tents according to the* *missions they receive from YHWH, and whose Oil is added to ours through the Name of the Father (the central shaft), as the Holy Spirit administers them, so that the seven lamps may be lit within us according to the will of YHWH.*

That is why it is only in the presence of the Name of the Father that coordination between the angel and the human is possible. The Name of the Father is present when you pray to the Father in the Name of Jesus, when you walk in obedience and sanctification, when you praise and worship YHWH, and when you read the Bible and meditate on what is written in it.

2. *The seven-branched lampstand is not very different from what the followers of certain religions proclaim when making the sign of the cross: in the Name of the Father, and of the Son, and of the Holy Spirit.*

   The seven-branched lampstand is, in a way, a sign of the cross: the portion of the Father in the middle, the portion of the Son and His descendants on one side, and the portion of the Spirit and His descendants on the other side.

3. *I dare to believe that just as humans master the material and visible world, so do the angels master the immaterial and invisible world; however, it is primarily from the immaterial that the material is managed.*

To put it more precisely, the 6 are not just 6 worlds; they are also 6 dimensions, 6 realities, 6 stages, 6 days (not necessarily days of 24 hours), and so on. Indeed, the lamps of the lampstand are the seat of our discernment; they reflect the degree of consciousness and visualization we have of the 6 dimensions created by YHWH. We can only truly understand things and become masters of the 6 if all 7 lamps are lit within us, and it is only through communion with the Father, through Jesus and the Holy Spirit (and His branches), that all 7 lamps are lit, making us living and animated statues of YHWH, the image of YHWH, to rule the world, as mentioned above.

> ***Then God said, "Let us make mankind in our image, in our likeness, so that they may rule..."*** *(Genesis 1:26)*.

As mentioned above, by themselves, rebellious humans who unite with demons can also light these lamps, but they can never light all 7 lamps; they can only reach 6, each bringing the light of 3 lamps, for the Bible says:

> "... ***The number of the beast is the number of a man. That number is 666*** *(Revelation 13:18)*."

Yes indeed, on either side of the central shaft of the lampstand, there were 6 branches—3 on the left and 3 on the right—and each branch bore 3 almond blossoms, that is, 3 fruits, along its length (Exodus 25:31–40).

When the spirit of a rebellious descendant of Jesus—that is, a human not converted to the Lord—unites with the spirit of a rebellious descendant of the Holy Spirit—that is, a demon—their union will never light 7 lamps, because one thing will always be missing: the perfection of YHWH, the central shaft with its lamp and its fruits.

Thus, the fruits of the spirit produced by the union of the rebellious human and the rebellious angel will give the light of 6 lamps coming from 6 branches, each bearing 3 fruits, and this light will amount only to 666: an imperfect light—imperfect to overcome temptation, imperfect to rule over sin, imperfect to live eternally; in short, a light which is darkness.

Thus, Jesus said:

> ... ***If then the light within you is darkness, how great is that darkness*** *(Matthew 6:23)****!***

666 is therefore the mark of all those who:

— have not embodied Jesus Christ and are not led by the Holy Spirit;

— allow themselves to open their third eye on their own and to look into or travel within the invisible realm: demons—beings often invisible even when the third eye is opened—will always be present, even without a formal invitation having been addressed to them; they will enter the holy place of the human who opens his third eye alone and will ally themselves with him in order to mislead him (bad decisions, wrong choices, addictions, false dreams, false visions, etc.).

— willingly enter into pacts with demons in order to see into the invisible realm or to move within it, or to secure their survival. These demons are also referred to as idols, gods, laws, spirits, and ancestors (1 Corinthians 10:19–22).

Never open the third eye by yourself...

As for you, simply receive Jesus Christ and remain in communion with the Holy Spirit, and you will have enough of the Oil of YHWH and the fire of His 7 lamps lit within you to see into the invisible and to know the inaccessible at the time appointed by Him and according to His good will, collaborating with His angels within you, to live a life according to His calling and His will for you.

- ***The secret things belong to the Lord our God, but the things revealed belong to us and to our children forever, that we may follow all the words of this law*** *(Deuteronomy 29:29)*.

- ***Call to Me and I will answer you and tell you great and unsearchable things you do not know** (Jeremiah 33:3).*

The prophets of YHWH, as well as those who open the third eye by themselves or through demons (enchanters and others), can therefore all have visions, dreams, and ideas, since they all possess a form of light to see into the invisible; but only YHWH can instantly see everything on the earth, in the heavens, in the seas and the waters, as well as in the burning cloud of the abyss.

Thus, only what YHWH speaks remains irreversible, true, and precise forever, for before His faces there is no shadowed area. Therefore, wait upon YHWH; do not look into the invisible by yourself, do not open the eye of the soul (the third eye) by yourself!

> ***Ask the Lord for rain in the springtime; it is the Lord who sends the thunderstorms. He gives showers of rain to all people, and plants of the field to everyone.***

> ***The idols speak deceitfully, diviners see visions that lie; they tell dreams that are false, they comfort in vain. Therefore the people wander like sheep oppressed for lack of a shepherd** (Zachariah 10:1-2).*

An addition...

Finally, if an addition were to be made concerning the 7-branched lampstand, I would say that one of the major components of the Oil of the central shaft is certainly Love (love your neighbor as yourself, and love YHWH above all, and worship Him alone... Matthew 22:36–40); and one of the most striking details is that the 6 other branches—3 on each side—each sprang forth just above a fruit-and-flower pair of the central shaft (Exodus 25:34–35).

Yes indeed, if the 7-branched lampstand symbolizes the spirit at work within a human being, and if each branch of the lampstand bears fruits—the fruits of the spirit that is within the human—by whose light the worlds created by YHWH are meant to be governed, then it is certainly through Love (for YHWH and for one's neighbor) that each fruit is meant to be expressed; Love should be the primary motivation behind the expression of every talent, every behavior, every gift received, and every human skill.

For where YHWH is not loved and actions are not carried out for His glory, and where one does not love one's neighbor as oneself, each of the 9 fruits of the spirit on the human side of the lampstand (3 on each of the 3 branches on the human side), and each of the 9 fruits of the spirit on the angelic side of the lampstand (3 on each of the 3 branches on the angelic side), becomes outright the opposite of what it was originally meant to be.

Thus, with reference to the 9 fruits of the Spirit mentioned in Galatians 5:22–23, without Love as defined above, love becomes hatred, joy becomes sorrow, peace becomes conflict, patience becomes impatience, kindness becomes wickedness, goodness becomes malice, faithfulness becomes unfaithfulness, and gentleness and self-control become brutality and lack of self-control.

“Love for YHWH (worshiping YHWH alone and giving glory to Him alone) and love for one’s neighbor (loving one’s neighbor as oneself)” is therefore certainly the perfection of YHWH—the fourth branch borne by every soul attached to Jesus Christ and by every angel of YHWH—and this is the common point that unites humans and angels to produce the light of 7 lamps.

True love for YHWH and for one’s neighbor can exist in a human only when that person has embodied Jesus Christ. A spirit detached from the Holy Spirit cannot be a spirit of Love, for it is written: Love is from God, and whoever loves is born of God (1 John 4:7).

Do you want to have Love within you? Then receive Jesus Christ and embody Him within you, and daily maintain communion with the Holy Spirit, for Love is the portion that YHWH alone can give to a being; it is not the fruit of mental effort, but rather the result of a new nature, a new state of the soul that can be given to a human only by YHWH alone through Jesus Christ, as it is written that *we are*

God's handiwork, created in Christ Jesus to do good works, which God prepared in advance for us to do (Ephesians 2:10).

> *If I speak the tongues of men or of angels, but do not have love, I'm only a resounding gong or a clanging cymbal.*
> *If I have the gift of prophecy and can fathom all mysteries and all knowledge, and if I have a faith that can move mountains, but do not have love, I'm nothing.*
> *If I give all I possess to the poor and give over my body to hardship so that I may boast, but do not have love, I gain nothing.*
>
> *Love is patient, love is kind, it does not envy, it does not boast, it is not proud.*
> *It does not dishonor others, it is not self-seeking, it is not easily angered, it keeps no record of wrong.*
> *Love does not delight in evil but rejoices with the truth.*
> *It always protects, always trusts, always hopes,*

always perseveres.

Love never fails... *(1 Corinthiens 13:1-8).*

- ***The bread of the Presence***

DIAGRAM 17: TABLE OF THE BREAD OF THE PRESENCE

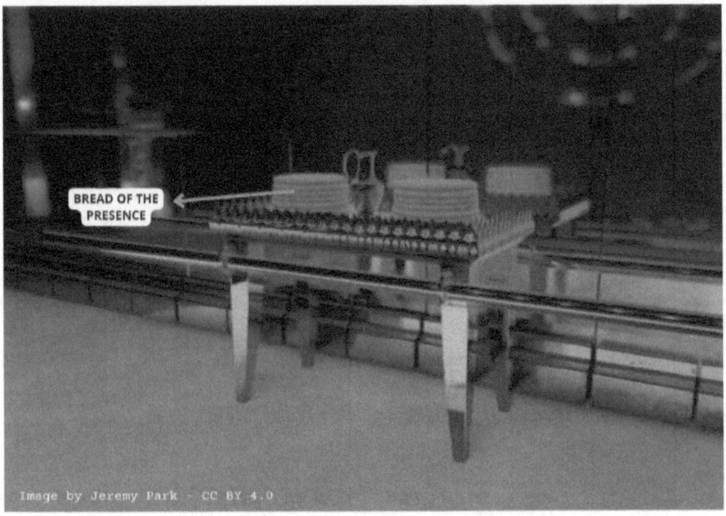

The loaves, in the first chamber of the tent, represent the Word of God (promises, guidance, teachings, good ideas for prospering, gifts from God, deposits arising from biblical meditations, revelations); these loaves are a reserve for our soul, enabling us to endure in the Christian walk, to make sound decisions in life, and to bring the enemy to nothing.

By the Spirit of YHWH within us, this reserve is also used as a sword against the enemy and adversity ***("Take... the sword of the Spirit, which is the Word of God"** – Ephesians 6:17)*, as well as for our comfort, our right conduct, and our salvation.

Only the one in whom all 7 lamps of the lampstand are lit **can see these loaves and make use of them**—that is, the one who has embodied Jesus Christ within himself and who walks in communion with the Spirit of YHWH. Thus it is written:

> ***Now we have received, not the spirit of the world, but the Spirit who is from God, that we might know the things that have been freely given to us by God*** *(1 Corinthians 2:12)* .

These loaves are certainly the daily bread spoken of by Jesus in The Lord's Prayer (Matthew 6:11), and they are also the seed planted by the Sower in the parable of the Sower (Matthew 13:1–23).

They are the nourishment of our spirit (spirit = the gold covering the walls and the objects of the Tabernacle): they give our spirit strength and understanding to cultivate our courtyard (to cultivate the soil of our soul and manage our affairs) and to lead us on the path of Life.

Certainly, YHWH gives daily bread to every human being and plants His seed in each one, for He is a good Father; yet many do not benefit from it, just as in the parable of the Sower:

- Some are by the wayside when the bread is given to them (no communion with the Spirit of YHWH, drunkenness, debauchery, distraction, etc.), and the demons, who are also spirits seeking food and who can enter the holy place, steal the bread and eat it.
  This is therefore the portion of bread that falls by the wayside and that the birds eat (Matthew 13:4, 19).

- Some receive the bread with a less persevering heart and give up very quickly when adversity comes; this is the portion of the seed that fell on rocky ground (Matthew 13:5–6, 20–21).

- Some have thorns in their holy place; that is, the horns of the altar of incense, inner scatterings that disturb their focus, and they cannot see, through their inner lamps, the bread that has been given to them.
  Jesus said that these thorns represent the worries of life and the deceitfulness of riches (Matthew 13:22), and this is exactly what the horns of the altar of incense are for the human soul (worries of life, worldly pleasures, addictions).

Finally, some, having embodied Jesus, living in communion with the Holy Spirit (through prayer, obedience to YHWH, praise, and worship), and being persevering, have all 7 lamps of their lampstand lit, and have no altars of incense nor their horns to disturb them (horns neutralized by the blood of Jesus). Thus, they can see the loaves that YHWH gives them and feed their spirits with them. This is the seed that falls on good soil, which makes the whole Tabernacle prosper and causes it to produce much fruit in its courtyard (Matthew 13:23).

Intruders in the human being...

Very often, tabernacles—that is, human beings—are invaded by intruders. These intruders are demons, who enter into humans either by themselves or through the intermediary of a *flying tent* (a wandering soul) that goes to operate in the courtyard of another person, carried by demons who assist its movement and inspire it, while in reality keeping it enslaved because of something it has eaten or used, over which the demon had claimed a right of ownership. This is what witchcraft is: consumables sacrificed to idols, to demons, in order to place them in communion with the human soul (1 Corinthians 10:19–20).

It is human sin that gives demons the authorization to sow disorder, for the sting of death is sin, and the power of sin is the Torah, the Law (1 Corinthians 15:56).

> ***Judah's sin is engraved with an iron tool, inscribed with a flint point, on the tablets of their hearts*** *(the tablets of the Law in the Ark of the Covenant)* ***and on the horns of their altars*** *(Jeremiah 17:1).*

However, **Jesus Christ** bore all our sins upon Himself in order to strip the demons of their right to enter into us, and He shed His blood to neutralize their horns; He wore the crown of thorns upon His head, as if to say that He took upon Himself all the worries, all the addictions, all horns of the altar of incense that can exist within a human being. Therefore, repent, and confess Jesus Christ and His work for you again and again, and you will have peace—the peace that YHWH alone can give.

Yes, Jesus Christ took away from demons—who act either on their own or through sorcerers whom they hold in bondage—the right to torment human beings, by bearing within Himself all the sins of humanity of all times (1 Peter 2:24). Thus, it is written:

> ***Awake, awake, Zion*** *(Zion = the tent, the human soul)*, ***clothe yourself with strength! Put on your garment of splendor, Jerusalem*** *(Jerusalem = the courtyard, the human body, the container of the soul)*, ***the holy city. The uncircumcised*** *(the witchcraft-practitioner's soul)* ***and the defiled*** *(the demon)* ***will not enter you again*** **(Isaiah 52:1).**

And Jesus Christ, by His death, took away from demons the right to keep human souls captive in witchcraft *(by making them wander and act wrongly because of something they had consumed)*, for it is from Jesus Christ that every atom of the universe comes, since all things were created by Him and for Him, and there is only one Lord, Jesus Christ, through whom are all things and through whom we exist **(Colossians 1:16; 1 Corinthians 8:6)**.

Jesus Christ therefore died and rose again to redeem human beings from the debt of every atom—every material or immaterial particle—that the enemy had claimed as his property. Therefore, confess again and again that everything you have already eaten or used, everything you eat or use now, and everything you will eat or use, is the body and the blood of Christ, He from whom every atom of the material and immaterial universe comes.

Do this again and again; eat and drink in remembrance of Him whatever is set before you, so that witchcraft may have no power over you!

> ***Jesus said to them, "Very truly I tell you, unless you eat the flesh of the Son of Man and drink His blood, you have no life in you.***
> ***Whoever eats My flesh and drinks My blood has eternal life, and I will raise them up at the last day.***
> ***For My flesh is real food and My blood is real drink.***
> ***Whoever eats My flesh and drinks My blood, remains in Me, and I in them*** *(John 6:53-56)*.

Through your repentance, by abiding in Jesus Christ, and by walking in sanctification and communion with the Holy Spirit—while never again looking into the invisible realm by yourself, and never again opening your third eye by yourself, if that was ever the case *(to open the third eye by yourself = to open directly the door of your Holy Place to the demon)*—the enemy no longer has any right over you. Therefore, you must do this again and again, for the enemy has lost every right over you, for eternity, in Jesus Christ!

> ***When you were dead in your sin and in the uncircumcision of your flesh, God made you alive with Christ. He forgave us all our sins,***
> ***Having cancelled the charge of our legal indebtedness, which stood against us and condemned us; He has taken it away, nailing it to the cross.***
> ***And having disarmed the power and authorities, He made a public spectacle of them, triumphing over them by the cross*** *(Colossians 2:13-15)*.

Another point is that on every Sabbath day, the priests replaced the old loaves with new, hot loaves (Leviticus 24:5–9). The fact that new loaves were brought every Sabbath (the seventh day) shows that everything YHWH gives is perfect; everything He gives belongs to the dimension of 7; everything He gives is a Sabbath—it requires no correction; everything YHWH gives remains perfect for eternity, because it is approved in the dimension of 7.

Also, in order to keep every bread that YHWH gives you (every instruction, every direction, every promise, etc.), it is important that you memorize it and do not forget it (write it down and read it often), and that you always say thank you for it; for by memorizing the bread and giving thanks for it, your soul feeds on it, and confessing your bread and giving thanks for it are a fire that drives away the stealing birds, the demons.

That is why pure incense always accompanied the bread of the Presence (Leviticus 24:7); this is also why the Word of God tells us not to let ourselves be distracted by false teachings, for losing faith in the bread that YHWH gives allows the enemy to plunder that bread.

***Beware lest anyone cheat you through philosophy and empty deceit, according to the tradition of men, according to the basic principles of the world, and not according to Christ** (Colossians 2:8, NKJV).*

*[Bonus: **You might still be wondering: Angel of God, Spirit of God, Jesus, humans—what exactly is the nuance?***

That is a good question! Indeed, just as those made of earth—that is, humans—are miniature versions of the Lamb ("I am the Vine; you are the branches" ... John 15:5), so too those made of breath—that is, angels—are certainly miniature versions of the Holy Breath (the Holy Spirit), for it is written that angels are spirits (Psalm 104:4).

And if Jesus said that we are His representation, giving the example of those who, in reality, visit or feed Jesus when they do it for the least of humans (Matthew 25:31–34), then the one who sees the Angel of God also sees the Spirit of God, for angels are also branches, offshoots of the Holy Spirit, just as humans are branches of Yeshua.

And it is through His branches, the angels, that the Spirit of YHWH the Father works in us, for our salvation and for the proper functioning of the Kingdom of YHWH, just as it is through humans who have received the nature of the Lamb—that is, Christians—that the Lamb (the Father's thinking Arm) acts, according to the will of YHWH the Father carried out by the Holy Spirit.

Yes, the angels act in us, which explains why the entire ceiling of the Tabernacle tent (as mentioned above) was made of 10 finely twisted linen coverings on which cherubim were embroidered (Exodus 26:1). *Thus, Zechariah, for example, speaks of the angel (the messenger) who speaks through him.*

> ***And behold the angel that spoke in me went forth...*** *(Zachariah 2:3, Douay-Rheims Version of the Bible)*

Indeed, only Jesus, forever, can be filled with the entirety of the Father's Spirit—that is, the totality of the Father's revealed thought, which Jesus perceives in real time, thanks to the full Holy Spirit dwelling in Him.

For Jesus Christ and the Spirit of God are the only two Trunks that take root in the Father (Zechariah 4); humans take root in Jesus, and angels in the Holy Spirit; humans and angels are therefore channels, branches of One or the Other. Thus, speaking of Jesus, John the Baptizer said:

> ***"For the One whom God has sent speaks the Word of God, for God gives the Spirit without limit*** *(John 3:34)* ***."***

The Trunk therefore harmonizes with the Trunk, and the branches harmonize with the branches. We and the angels are cousins.

Thus, the Holy Spirit is a Person—that is, a complete being with full consciousness; He is a Tree with full consciousness: the Trunk possesses full consciousness, and the branches that come from Him and depend on Him (the angels) also have full consciousness.
It is somewhat like us humans: Jesus has His full consciousness, and we also have ours.]

There is still so much to say about the Tabernacle... Do you want to learn a bit more about, for example, the Garden of Eden and the Tree of the Knowledge of Good and Evil? If so, I invite you to consult the book by the same author entitled: *70 Centuries to Go for the Universe: The Reparation.*

This was the creation of the human being: on the sixth day, YHWH the Father made, from the earth and branches of the Lamb, tabernacles (tents with their enclosures, that is, souls with their bodies). Then He breathed a breath of life into the tent: He lit the 7 lamps of the lampstand with His Oil, channeled through the Lamb as well as the Holy Spirit and His branches. Thus, the tent became a living tent; Adam (the being made of earth, the human) became a living soul, a living statue of YHWH, able to live for eternity.

And remember, the lampstand was entirely made of gold, which symbolizes the spirit. Seven lamps being lit means that to the Oil of the Son (the Son's own Spirit) dwelling in the human is added the Oil of the Holy Spirit, channeled through His branches, and the two Oils are cemented by the Name of the Father—the fourth branch—and His Oil, which is certainly Love, carried by each of them. Thus, by himself, a human cannot accomplish what YHWH truly expects of him.

> ***But it is the Spirit in a person, the breath of the Almighty, that gives them understanding*** *(Job 32:8)*.

Just as the Lamb and the Holy Spirit are eternally united in their very nature, humans and spirits are likewise united in their very nature. Thus, wherever human souls are present, there will always also be spirits—angels and/or demons.

> *[Something never to be done!*
> *The relationship between a human and the Angel of YHWH is a one-way relationship: the human speaks to YHWH in the Name of Jesus, and through his angel (his messenger), often in various forms—dreams, visions, thoughts, etc.—YHWH reveals His thoughts and gifts to the human.*
>
> *And those who have had experiences with the true Angels of YHWH know that the angels of YHWH are discreet; one does not retain their faces—they appear as strangers (often unknown humans) in our dreams and visions.*
>
> *Therefore, do not pray to angels, worship angels, call upon them, or seek them out*

yourself to speak to them directly! In fact, it will not even work; and if it seems to work, or if someone tells you they have done it successfully, then it was not an angel of God in that matter, but rather a demon.

The true angels of YHWH are not independent workers; ***they do only what YHWH commands them****, and they are not gods. They are our companions in the work of God's Kingdom* ***(Revelation 22:9)****.*

It is the demons (the contaminated spirits) who began to call themselves "elohim" (gods), ancestors, spirits, laws, etc., and even to demand worship from humans. Unfortunately for them, they are not complete like our Elohim, their Creator and ours; they lack Material Substance and rely on humans to carve for them a physical form—a idol, a tool—made from the material dust of the Lamb.

Worship is reserved for the Father and His two Trunks (Jesus and the Holy Spirit) alone—that is, for YHWH alone; do not worship a branch!

Indeed, YHWH holds the Creative Substance, His Thinking Arm, Yeshua, whose dust brought forth the entire universe with its trillions of galaxies. By what physical form could one represent our God? He is infinite, and everything comes from Him.

So what should you do?
You, speak to YHWH in the Name of Jesus, and let the angels do their work according to the orders they have received from Him.
You speak to YHWH, and He speaks to you through the Holy Spirit via His branches (His angels, His messengers); do you want answers from YHWH? Then speak to YHWH, not to His angel!
Be careful, and do not fall into the trap!]

YHWH had thus built for Himself a beautiful tabernacle—the human being. However, an event occurred, and YHWH's tabernacle, which had once been a beautiful garden inhabited by His Spirit, was left desolate and destroyed.

Humanity and the world were now drawn into countless wars, when, in a time known as the era of the prophets (eighth–seventh century before the birth of Yeshua), YHWH announced that He would return to the tabernacle so that it might once again be inhabited, rebuilt, and replanted, forever.

And YHWH promised that, from the time of the announcement of His return to the tabernacle, there would remain only **70 CENTURIES to go for the universe**.

END OF VOLUME ONE